ENVIRONMENT, ENERGY, HEALTH

ENVIRONMENT, ENERGY, HEALTH

Planning for Conservation

Prof. V. Vidyanath

GYAN PUBLISHING HOUSE

NEW DELHI-110002

ENVIRONMENT, ENERGY, HEALTH
Planning for Conservation
(*Health*)

ISBN : 81-212-0711-8

Published in 2000 in India by
Gyan Publishing House,
5, Ansari Road, Darya Ganj
New Delhi-110 002

Phone: 3282060, 3261060 Fax: (011)-3285914
E-mail: gyanbook@del2.vsnl.net.in
Visit Our Web Site At http://www.gyanbooks.com

Laser Typesetting at A.S. Printers, Delhi

Printed at Mehra Offset Press, Delhi

CONTENTS

PART—II

Alternate Energy Sources

PART—III

Diet and Heath Interrelationships

FOREWORD

The rapid industrialization and mindless consumerism and change in lifestyles of men for generating and spending of wealth during the second half of the twentieth century has created an unprecedented problem of environmental degradation. The delicate balance between man and biosphere is getting disturbed and if continued further may reach to a point of no return which may mean doomsday for the survival of man on this beautiful planet earth. This being a serious problem, it has been attracting attention of scientists, technologists, policy makers, environmentalists and general public all over the world. Governments, international agencies and non-governmental agencies have been deliberating and suggesting measures to be taken to preserve the environment. This book is an attempt to bring awareness to all concerned and the general public of the seriousness of the problem.

This book consists of articles on three inter-related aspects of environment, energy and health. In the First Section the articles deal exhaustively on various aspects of land, air and water pollution with several suggestions for preserving the environment. The Second Section deals with the use of alternative energy resources to agument and supplement to meet the future energy crisis, which is going to occur due to unprecedented exploitation of fuel wood and fossil fuel resources. The Third Section deals with articles on health, which are related to preservation of environment.

The author of the book who has been a Professor of Geography in Osmania University is a distinguished teacher, researcher and a prolific writer. Due to his abiding interest in human well being and concern for improving the quality of life without disturbing the delicate balance between man and biosphere, he has published many articles on this topic in the national newspapers. He has also contributed well for the programmes on Television under the EMRC of University Grants Commission. I am confident that the clarity and simplicity of his writing will appeal to the students and general readers as well.

Prof. T. Navaneeth Rao

PREFACE

Of late there has been much awareness in both developed and developing countries for environmental preservation and ecosystem. With rapid growing population in developing countries like India and with accelerated technological development in developed countries in particular, there is so much environmental degradation and ecological imbalance. Environmental degradation is not a new phenomenon. It has started with the primitive man in early ages. Even though the exploitation of natural resources started with the primitive man, he could not do so much since he was at the mercy of the environment on account of his poor knowledge to exploit nature. In the beginning, man was a fruit gatherer living on fruits and raw vegetables. Later he learnt the hunting of animals and eating raw meat and subsequently, i.e., about 1,50,000 years ago, he learnt the use of wood as a source of fuel energy for giving warmth when it was cold, for seeking protectiong from wild animals and for preparing more digestable food by way of cooking. Since then, he started the exploitation of nature by burning wood. But with the advancement of man's knowledge in the field of science and technology the process of environmental degradation has been accelerated. His exploitation of nature has started in two directions, i.e., the exploitation of nature's space and exploitation of nature's resources.

Environment in simple terms means surroundings. The term environment originatres from the French word *Environir* (to surround) and it denotes all the conditions, circumstances, and influences surrounding and affecting and organism or group of organisms.

There are broadly two types of environments over the planet Earth, one is the physical/abiotic environment comprising land, air and water and other is the biotic environment consisting of plant and animals including man, which is also called biosphere. The interactions and interrelationships between and within non living abiotic and biotic environments are expressed under the term 'Ecosystem'. In other words the study of an ecosystem is primarily concerned with the structure (composition) and functioning of a living system with a non living system. Even within the living and non living systems interactions take place. Depending upon the area of operation of various elements belonging to biotic nd abiotic environments, different ecosystems have come into existence, each with its own distinct characteristics. For example, a coastal ecosystem is different from a lake ecosystem, and so on.

In an ideal ecosystem an equilibrium or stability is maintained between these two environments, which is called the ecological balance. Often this ecological balance is disturbed due to over interaction and dominance of man of biotic community, who exploits nature to satisfy his wants and desires causing environmental degradation. In fact the damage caused by certain natural/abiotic agents like earthquakes, volcanic eruptions, floods, cyclones is much less when compared to that of man.

The land, air, water—the three components of abiotic environment, have been very much polluted and various ecosystems have been largely affected due to over exploitation of resources of land and water with innumerable human activities like unscientific cultivation practices, for example, shifting cultivation, industries, transport, construction and over grazing of animals and certain social religious customs and practices. The land, air, water have been polluted with industrial wastes, including hazardous wastes; industrial effluents; urban sewage and garbage; pollutant gases like sulphur dioxide, nitrous oxide, carbon dioxide and carbon monomide by burning fossil fuels like coal and petroleum, large scale emission of sulphur dioxide, nitrous oxide; gas emissions from industries, causing acid rains and methane gas from paddy fields. All the green house gases (carbon dioxide, nitrous oxide methane, troposphere ozone are causing a rise of the temperature of earth's surface resulting in the rise in sea-level. The chlorofluoro carbon chemicals (CFCs) present in troposophere are also responsible for rise of temperature by absorbing heat. Fifty per cent of the green house effect is caused by carbon dioxide alone. Among these green house gases, there is significant rise of carbon dioxide concentration during the last century due to burning of petrol and diesel oils in transport vehicles, and coal in industries.

The large scale destruction of forests which act as carbon sinks has also contributed to this enormous rise of carbon dioxide. The rapid increase of population particularly in developing countries has also contributed by exhaling carbon dioxide in respiration phenomenon. At present the carbon dioxide is rising by 0.5 per cent per annum.

The rise of surface temperature is warming up sea water which expands, causing rise in sea level. As a result of this phenomenon it is expected that several coastal areas, and some of the islands like Maldives, are likely to submerge in the next few decades.

Another alarming effect of atmospheric pollution is the large scale depletion of ozone, the life saving gas from the atmosphere due to presence of certain chemical substances known as chloroflurocarbons (CFCs) emitted by certain industries like refrigeration, polythene and foam rubber.

Even the biosphere has not been spared by man. Forests are being destroyed for energy needs for certain industries like paper; and for other activities like construction, and cultivation. Agricultural lands have been converted for urban expansion around large cities in many countries of the world. Large scale destruction of Brazilian forests has taken place for feeding the cattle to export meat to USA.

Many estuarine and marine species like fish and others have been perishing due to the presence of certain chemical substances of fertilizers, pesticides found in the drained water from agricultural fields entering into lakes, rivers. Marine species are also affected by toxic hazardous wastes dumped along the coasts.

Over exploitation of round water has lowered the water table and even caused the sinking of land surface is some parts of the cities like Bangkok, Tokyo, Mexico and Houston.

Social habits and religious customs and practices have also contributed a lot to water pollution. River Ganges has become the most polluted river with leftover waste material of religious rituals, half burnt corpses thrown in the river, etc. Immersion of coloured idols in rivers and tanks is anothr religious custom polluting the water Deforestation along canal, river bunds and along beaches, as a social habit is another example of water pollution. Development of slums and sckatter settlements, traffic congestion, noise pollution in cities are due to large scale migration of people from small cities and towns and rural areas in search of employment have affected the urban ecosystem of several cities.

This book consists of articles on three interrelated aspects of environment, energy and health. As such it has beer divided into three sections.

In the first section there are articles which have been dealt exhaustively on various aspects of land, air, watr pollution and at the end some remedies have been suggested to preserve environment.

The second section dealth with the alternature/supplementary energy resource potentialities and utilization and the possibilities for further utilization to meet the future energy crisis which is likely to occur in the next few decades on account of present overexploitation of fuel wood and fossil fuel resources. There is already sufficient progress in utilizing the biomass for generating biogas, solar and wind energy under non conventional energy resources in our country. A detailed account of the development of non conventional energy resources has been included in this section. The prospects of utilization of coal methane gas, and geothermal energy in some parts of our country to supplement the conventional energy have been also dealt with in this section.

The third section deals with a few articles on health. In the article on 'Vegetarianism—an ideal choice', it has been mentioned how vegetarian diet, appart from being more healthy, is environment protective, and also provides food to a large number of poor people who can't afford a non-vegetarian diet and depend upon grains. The significance of dietary fibre to keep a man healthy has been dealt in a separate article. Physical fitness is an essential requisite to keep a man engaged in socio-economic activities. As such two articles one dealing with walking exercise and its health benefits and another entitled 'know your health risk by body weight' have been included under this section.

Environmental preservation is a must for the survival of the entire present and future biotic community.

For this measures are being taken to promote pollution free technology for industries and transport vehicles; reducing the number of transport vehicles by providing houses for commuters near working place; increasing the public transport; decentralisation and shifting of industries and other institutions from large cities to medium and small towns and growth centres and also establishing new ones in these centres; checking of large-scale migration to big cities; determining the optimum carrying capacity of a city with reference to pollution by taking certain environmental parameters like drinking water availability, green belt converage and carbon dioxide concentration to check further expansion and to preserve urban ecosystem and lastly bringing about change of outlook and lifestyle of the people. A chapter on Environment and Constitutional protection includes the importance of understanding the seriousness of the problem by the people and their awakening through mass media. Environment laws should be strictly enforced. There are enough number of articles in our Constitution of India to protect environment.

It is necessary to plan our interactions with nature and regularise our impact on nature. The objectives of ecodevelopment is the creation of environment that sustains life but also enriches life, harmonising the works of man and nature for greater goodwill of all. It does not mean that man does not touch or interfere for the deformation of environment since it is inevitable for the growing culture of society. The inevitable degeneration of environment should be a temporary phenomenon and it should be regenerated quickly. The essentiality of man and nature is based on man as natural being, nature as human society. The objective reality is the mechanism of interactions of social man and non human nature. The naturalisation of man and humanisation of nature proceed simultaneously through interaction of man with nature. Since the interaction of man with nature is inevitable the basic objectives of development technologies for environment preservation and sustainable development should aim to achieve quick and rapid regeneration of environment wherever and whenever a man disturbs the ecological balance and finally maintain harmony between environment and man's socio-economic needs.

Since the adverse impact of environmental degradation does not confine to the place of origin and spreads to other parts. it is considered as a global phenomenon. Sometimes the efect is not felt at the place of origin but at far off and distant places as noticed in the case of acid rains which occured in European countries, due to emissions of sulphur dioxide from industries located in England. As such one should think globally before acting locally.

The world as one community we have share the earth's resources by cultivating universal brotherhood. One should bear in mind that what Gandhiji had said, "The earth has given enough to satisfy everybody's wants, and not to satisfy everyone's greed". therefore one should optimise satisfaction with minimum wants instead of maximising output and over exploitation of resources to ever increasing wants which result in environmental degradation.

I am grateful to all the participants in my endeavur to present this book to the reader. Hope it succeeds in bring out the creation of a sustainable environment process.

V. Vidyanath

PART—I
LAND, AIR AND WATER POLLUTION: DIFFERENT ASPECTS

1

ENVIRONMENT AND CONSTITUTIONAL PROTECTION

Of late there is so much concern about environmental degradation and remedial measures to be taken to protect it. In this regard a very big conference was held in Rio de Jeneiro in Brazil a few years back and several Nations participated in the conference. Under Agenda 21 of this conference a number of remedial measures to protect the environment have been accepted by all the Nations.

In all such conferences normally much is said about land, air, water pollution caused mostly by deforestation, industries, transport and the remedial measures to control the environmental degradation. But little importance is given to domestic pollution caused by an individual or a group of family members who adopt a certain lifestyle and outlook. Unless there is a change in the lifestyle and outlook of people it is not possible to preserve environment. What an individual does in his house, has a global effect. As such, one should think globally and act locally. Such an environmental awareness has to be created in rural communities as well as educated urban affluent societies. Educating even the educated is must so far environment is concerned, since environmental pollution caused by affluent urban societies is much more than that of poor rural illiterate folk.

In the democractic system of our country much emphasis has been given for protection of environment through a few important articles incorporated in our Constitution.

There is a misconception even among educated middle-class and urban affluent societies, that everyone has absolute freedom and fundamental right to live as he likes by spending as much as he could, to have a luxurious life by over-exhausting natural resources like food, fossil, fuels which may lead to not only over-exhaustion of resources, but also cause environmental pollution.

An examination of a few articles of our Constitution would dispel this impression. It is true that our Constitution gives lot of freedom, like freedom of expression, freedom of speech, worship, freedom to spend money for living for all citizens of our country and can have freedom to have their own customs and habits. But one can enjoy these privileges so long as they do not adversely affect him, his health, or society in which he is a member. But many misuse these privileges of freedom guaranteed to an individual under Article 19 which deals with fundamental rights in our Constitution. To check environmental pollution by misusing this article a few articles have been also included in our Constitution to protect environment from pollution.

For example, Article 32 which deals with public interest litigation popularly known as PIL has been receiving wide public attention nowadays. This Article includes wider applications also, including environmental protection under ecology. Any citizen of our country can file a writ in the court of law against an individual or group of individuals or an institute or government to prevent environmental degradation/pollution in the interest and welfare of the society including himself.

Environmental degradation can be challenged in the court of law by any individual under Article 21 which ensures protection to life since environmental pollution is injurious to one's health.

Another safeguard through Article 48A deals with protection and safeguarding forests and wild life. It is the responsibility of the government to see that environment is absolutely protected from any agent. The government has got the responsibility to protect health of the people under Article 47. Therefore any damage caused to one's health due to environmental pollution or due to one's own lifestyle or pattern of living could be legally prevented under the Article. But it all depends upon the level of damage caused to an individual. Narcotic drugs have been banned all over the world since they affect the nervous system which may lead one to become insane, others like chewing tobacco, smoking of cigarettes, eating *gutka* and *pan masalas* are also equally injurious to health but their adverse effect on health is more slow. It has been established by medical scientists that cigarette smoking is susceptible to heart attacks, and chewing tobacco, eating *gutka* and *pan masalas* are likely to cause cancer. If the situation warrants government can ban them under Article 47 of our Constitution.

It has also been established that production of beef and meat is also indirectly causing environmental degradation. Large-scale distribution of grazing lands and forests has been taking place for feeding cattle, goats and sheep for the production of beef and meat. According to Menaka Gandhi 'a sheep or goat destroys 20 hectares of grazing land in its life time of five years and it costs Rs. 25000 to restore the degraded land. Due to this the lower Sivalik hills are becoming barren and streams coming to Haryana are drying due to grazing of goats. A vegetarian can subsist just on 0.72 hectares of land a year while 1.63 hectares of land are needed to produce 1 kg of meat. More hungry people of the world can be fed on grains if beef and meat production is reduced. Even from health point of view vegetarian food which consists of a lot of edible fibre is essential for health and is less contaminated. On the other hand non-vegetarian food has no fibre and is contaminated more with toxins.

As such the state has a right to ban any items of food for protecting one's health and environment as and when necessity arises by using the Articles 47 and 48 A.

It is also the fundamental duty of every citizen of India to protect and improve natural environment including forests, lakes rivers, and wild life under Article 51A (G) under fundamental duties. Any one who violates could be penalised.

From all these Articles it is clear that no one has a right to overspend money to have luxurious lifestyle which leads to the environmental degradation and also affects his own health and that of community.

It is the responsibility of the government to create an environmental awareness in both rural and urban communities and also enlighten them of the limitations of individual freedom.

One can enjoy fundamental rights under Article 19 of our Constitution in our democratic system. But one can enjoy these privileges if there is no adverse impact on society. Using mikes, causing noise pollution, throwing half-burnt corpses in Ganges river, over consumption of petrol causing air pollution, etc., can be cited as few examples.

One should bear in mind and follow what Gandhiji has said:

> The earth provides enough to satisfy every man's need and not every man's greed. Civilisation in the real sense of the term consists of not in multiplication of wants but in the deliberate reduction of wants.

Therefore one should optimise satisfaction by minimising output and over exploitation of resources for ever increasing wants. One should green his mind by cultivating, eco-friendly activities/hobbies like reading, writing, listening music, playing games, etc.

Awareness about the need to preserve environment for the welfare of present and future generations is a must and should be created in both urban and rural communities. Any one who is directly or indirectly responsible for environmental degradation is considered to be an environmental criminal and he should be penalised and punished. There should be strong political will to take up such measures provided law makers are not law breakers.

2

GLOBAL WARMING: CAUSES AND CONSEQUENCES

The atmosphere over the planet earth keeps the earth warm maintaining an average temperature of 15°C which is required for the survival of the entire biosphere. In the absence of atmosphare, the earth would have cooled to –20°C and the day temperatures would have gone beyond 100°C. But the atmosphere, acts like a protective shield or blanket preventing the heat radiation from the sun. This in the form of infrared rays of short wavelengths can penetrate through nitrogen and oxygen, the chief components of atmosphere with 20 per cent and 80 per cent composition respectively, due to the transparency of these gases. But the outgoing heat in the form of infrared rays from the earth's surface which is radiated in short wavelengths is absorbed by trace gases notably carbon dioxide (CO_2) and water vapour thus keeping the earth warm and habitable. This heat protecting effect, like an umbrella, is called the greenhouse effect. In a green house, i.e., a glass house used for growing plants in cold countries, solar heat radiation enters through the glass walls which are transparent to this type (short wavelengths) infrared radiation. The interior of the green house warms up and in turn radiates heat in longer wave lengths. The glass walls are less transparent to this type of heat radiation, with the result the interior of the green house has higher temperatures as it is heated up to an appreciable extent relative to its surroundings. A similar effect is noticed in a car when glass doors are closed. The gases responsible for green house effect are called green house gases. If green house gases, were not there the earth would have been so cold and unable to sustain life. The green house gases are carbon dioxide (CO_2) nitrous oxide (N_2O), Methane (CH_4), Chlorofluorocarbons (particularly CFC_{11} and CFC_{12}, ozone, and water vapour. Among these CO_2 is the principal green house gas.

With the development of science and technology, the human and economic activities like industry, transport have increased. As a result of these activities, there is a significant rise in the level of CO_2 concentration in the atmosphere, which is the principal and major green house gas and this is chiefly produced by burning fossil fuels like coal and petroleum.

The other green house gases are also present in considerable proportions in the atmosphere which also contribute for the rise of temperature. The origin and rise of these gases are due to significant development of industries, transport and agriculture, and power development chiefly. Emission of green house gases during the last century seems to have raised from 0.05 to 0.8°C. The rise of global temperatures has tremendous effect on climate, sea level, terrestrial, and marine ecosystems.

Green House Gases

Carbon dioxide

As mentioned already, carbon dioxide is the most important gas among all the green house gases since its contribution to global warming is higher when compared to all other gases. About 50 per cent of green house effect is caused by this gas only. Natural processes produce vast amount of this gas, terrestrial life emits about 100,000 million tons of CO_2 a year simply by breathing, while decomposing vegetation releases another 2000 to 5000 tons annually. Natural photosynthesis takes up the some amount of release from these processes. Human activities are upsetting this balance. Firstly they release extra carbon dioxide by burning fossil fuels for industries and transport. Then they upset the balance by the destruction of forests and other vegetation. Natural processes circulate carbon through the atmosphere, oceans and biosphere in what is known as geo carbon cycle. Human activities disturb this cycle by injecting CO_2 into the atmosphere. It has been estimated that between 40 and 60 per cent of CO_2 emitted into the atmosphere remains air born, the rest is taken by natural sinks particularly the oceans. As mentioned already, increase of CO_2 is chiefly due to industrial activity, transport and population growth.

Burning fossil fuels is the main source of man made CO_2 emissions. It is estimated to release 5000 million tons of carbon per year. Deforestation is the next important source. A recent study reveals that deforestation in the tropics may be responsible for a release of 310 to1300 million tons of carbon per year while conversion of forest soil to other uses, causes the release of another 110 to 250 million tons. In all man made changes including forest destruction, the effects of acidic deposition, and the destruction of grass land and other vegetation cover through desertification result in a net annual increase of 16 million tons. Deforestation in Brazil, Indonesia, Colombia, Thailand, Phillippines, Cote–de, Ivoire, Zaire, Peru, Equador and Mexico accounts for three quarters of these carbon dioxide releases.

Estimates of how much carbon dioxide will be emitted in future vary widely with different projections for fossil fuel use and different scenario for the destruction of earths vegetation particularly forests.

It has been estimated that CO_2 emissions may range between 6000 mil tons to 10000 mil tons by the year 2000 rising to between 12000 mil tons to 24000 mil tons by 2050. Release from destruction of vegetation may be at 1000 mil tons of carbon a year till the year 2000 but may rise to 7500 mil tons by 2025.

The concentration of CO_2 in the atmosphere varied over the ages. Global records suggest that it may have been about 200 parts per million by volume (PPMV) during the last ice age some 18000 years ago. Estimates of CO_2 in pre-industrial times range 250 to 290 PPMV. Recent analyses reveal that atmosphere concentration around the year 1750 was 280 PPMV. Precise and continuous measurements of CO_2 concentration have been made in Mauna Loa Observatory, Hawaii since 1958 clearly show that it has increased from 315 PPMV in 1958 to 345 PPMV in 1985. The mean concentration of CO_2 has been estimated to range from 494 PPMV to 627 PPMV by the year 2050.

The amount of CO_2 in the atmosphere is currently increasing by 0.5 per cent per year. It means on an average one ton of carbon for every person is put into the atmosphere each year by burning fossil fuels. About half of it is eventually used by the plants or used by ocean but the rest stays in the atmosphere.

More than 80 per cent of CO_2 emissions are caused by industrial activity by burning fossil fuels and cement manufacturing, during limestone processing. Industrial emissions of CO_2 have more than tripled in the fast 40 years, prompting fears of increased global warming. CO_2 accounts to 50 per cent of green house gases responsible for trapping heat inside the atmosphere. The annual per person emission had to be at or below 1.69 metric tons based on 1990 World population of 5.3 billion to prevent additional CO_2 build up in the atmosphere.

If atmospheric concentration of CO_2 rises by 50 per cent from current levels, average global temperatures rise by 1.5°C.

Region-wise distribution of industrial CO_2 emissions in 1950 and 1990 are shown in the table below:

REGION-WISE DISTRIBUTION OF INDUSTRIAL CO_2 EMISSIONS

Country	*1950*	*1990*
USA and Canada	47%	26%
China	1%	11%
Western Europe	24%	14%
USSR and Eastern Europe	18%	23%
Others	7%	20%

From the above table it is clearly seen that while the developed countries reduced CO_2 emissions, the developing countries increased CO_2 emissions with industrial growth.

Recent findings indicate that the emission from burning fossil fuels has been continuously increasing. The highest levels of CO_2 were recorded in the Asia Pacific region which included India, China and the Middle East according to London based World Energy Commission Study.

Researchers say that the Asia Pacific region, except Japan, Australia, and New Zealand has shown a 37 per cent increase in emission during 1990 as against the level of emissions in

Eastern Europe that has shown a gradual decline of CO_2 emissions. The Asia Pacific region stands at 1.5 billion tons a year which is 6.4 per cent higher than the figures in 1990 and 2.7 per cent higher than the CO_2 emissions in 1995. Other contributors to global pollution are USA, Germany, Denmark, and UK. United States which has only 4.73 per cent of world population contributes 22 per cent of CO_2 to the atmosphere.

It is clear that industrialised countries hitherto accounted 85 per cent of total carbon emissions. The developing countries have become significant only in the past two decades. CO_2 levels now are 25 per cent higher than they were in 1960.

Today 5 billion tons are added to the atmosphere due to human activity apart from process of breathing.

The greatest potential for reducing CO_2 emissions is through afforestation, according to N. Khanna and S. Gupta of Tata Research Institute. Their calculation is that 1870 mil tons of CO_2 can be reduced in 35 years.

Nitrous Oxide

Nitrous oxide (N_2O) emissions result from fertilizer use, changing land use, fossil fuel combustion and biomass burning. The amount of total emissions of N_2O have been estimated at 30 mil tons, a quarter of this is anthropogenic. The major sink for N_2O is reaction with activated oxygen in the stratosphere. Measurements for N_2O emissions show that it has increased from 289 PPMV in 1970 to 340 PPMV in 1985 and it is rising by about 0.2 to 0.3 per cent per year. This comes primarily from combustion. It is uncertain whether the increased use of nitrogen fertilizers in agriculture or increases in deforestation and other land use changes contribute significantly. However, another study (Scope Report No. 29, 1986) considers that applying fertilizers enhances the flux of nitrous oxide to the fertilizers. The study estimates that emissions of N_2O due to fertilizers to be 600-2300 tons a year and the increases from cultivated land to be 200-600 tons a year. By one estimate, the N_2O concentration in the atmosphere might reach 375 PPMV in the year 2050. Another study estimates that it might range from 392 PPMV to 446 PPMV. Nitrous oxide contributes 4 per cent to global warming. It is extremely stable once it gets into the atmosphere on an average of 150 years. While it is in the lower part of the atmosphere it absorbs heat but it can rise to stratosphere where it helps to destroy ozone.

Methane

Methane is produced by bacteria in wet and marshy lands, paddy fields, cattle dung, in the gut of termites and other wood consuming insects, biomass burning and garbage land fills and coal mines and natural gas leaks. The total amount in the atmosphere is probably between 400 and 600 mil tons. Natural wetland ecosystems represent a large source of gas, estimated annual fluxes ranging from 100 to 150 mil tons a year.

Methane emissions from wetlands vary depending upon soil, air temperature and soil moisture, the amount and composition of organic substrate and vegetation. The extensive organic rich arctic and boral wetlands are especially important sources of the gas and account for about half of the total world wide emissions from natural wetlands. Human activities that alter wetlands such as draining or flooding also alter methane emission. Rice paddies also produce the gas, estimates range from 30-170 mil tons a year. Altering the productivity of rice paddies or extending their area increases methane emission. A recent study estimates that methane emission may have risen from 75 mil tons a year in 1950 to 115 mil tons a year in 1980, because of increase in the extent of paddy fields. Domestic animals are estimated to produce 74 mil tons a year and termites between 15 and 150 mil tons.

Analysis indicates that methane concentration in the atmosphere was about 0.7 PPMV before 1850. In 1977 the average global concentration was about 1.52 PPMV and it reached 1.7 PPMV in 1985. The average global increase in methane concentration has been about 1 per cent in 1965 which means it has been rising very much faster than CO_2 concentration. Concentrations vary regionally. They are always 7 per cent greater in north temperate zone than in the south temperate zone because the major methane sources are predominantly located in the northern hemisphere. It has been estimated that methane concentration in the atmosphere might reach 2.3 PPMV globally in the year 2030. Another study estimates that it may reach 3.5 PPMV in the year 2050.

Like CO_2, it contains carbon and increases twice as fast as CO_2 in the atmosphere. Scientists are concerned that increasing concentration of this gas may make it more difficult for the atmosphere to decompose other green house gases such as CFCs.

According to an energy expert in Delhi, a minor reduction in methane emissions can be achieved by improving natural gas transmissions in western countries. Methane emissions vary from field to field depending upon moisture, type of soil, light, crop rotation, fertilizer use, and irrigation practices according to a German expert.

According to Anil Agarwal and Sunita Narain of Centre for Science and Environment, a Delhi based Non government organisation 'emissions from wet rice cultivation in India would range from 3 to 9 mil tons much lower than the World Research Institutes estimate of 18 million tons. Methane emissions from paddy fields can be reduced by using quicker yielding varieties of seeds'.

Chlorofluorocarbons (CFCs)

Chlorofluorocarbons particularly CFC_{11} and CFC_{12} have been emitting from industrial sources to the atmosphere from the past half a century. The chief sources of CFCs are aerosol (spray propellants), refrigeration and airconditioning, plastic foams, industrial solvents, sterelants for medical supplies. The annual emissions of these two main CFCs is estimated to be about 0.4 tons. Photochemical destruction mainly in the stratosphere, and very slow uptake by the oceans are only known significant sinks for CFCs. Concentrations have been rising very rapidly. In 1977 there were about 150 parts per trillion by volume (PPTV) of CFC_{11} in the atmosphere. It rose to about 226 PPTV in 1986. The concentration of CFC_{12} rose from 260 PPTV in 1977 to 392 PPTV in 1986. On an average CFC_{11} has been increasing by 4 per cent a year and CFC_{12} has been increasing 4.3 per cent per year. It has been estimated that CFC_{11} might reach 1100 PPTV in the year 2030 with CFC_{12} reaching 1800 PPTV. Another study estimates that the concentration of CFC_{11} would rise from 1379 to 2897 and CFC_{12} from 2359 to 3829 by 2050. However, the Montreal protocol which came into force at the beginning of

1989 observed that the concentration of CFC_{11} and CFC_{12} should be about 700 PPTV and 1400 PPTV respectively.

These long standing chemicals not only destroy the ozone layer in the stratosphere but are also powerful green house gases when they are in the lower layer of the atmosphere, i.e., troposphere.

USA produces 35 per cent of CFCs and another 35 per cent is produced by Western Europe. In the London conference it was decided to bring about a 50 per cent cut by 1995 and 85 per cent by 1997 and a complete elimination by the year 2000. CFCs have a global warming potential of 4500 to 7100 in 20 years-while CO_2 has sinks in forests and oceans, CFCs have no sinks.

When compared to US, India has a much lower production of CFCs being well below 0.0088 kg, its per capita consumption in the production of refrigerators and air-conditioners being proportionately lower.

Ozone

Ozone is present both in the stratosphere and troposphere of the earth's atmosphere. Troposphere ozone partly comes from stratosphere and partly arises from photochemical production involving nitrogen and methane and other hydrocarbons. The concentration of ozone in the troposphere was about 0.01-0.1 PPMV and appears to be increasing.

Ozone contribution to green house effect is complicated by interaction with trace "reactive" gases. Many of these gases are produced brologically or by burning biomass and the concentration of many are currently increasing in the atmosphere. Many do not absorb infrafed radiation and so are not green house gases in the strict sense, but they interact in ways that cause the production or destruction of other gases and can therefore indirectly influence green house effect. Ozone is also a byproduct of fossil fuel burning.

Water vapour

The amount of water vapour depends upon temperature. Warm air can store more moisture than cool air-water vapour absorbs heat very efficiently but it can give rise to the green house effect by itself. However, it serves to amplify the effects of other gases.

Consequences of Green House Effect and Global Warming

Recent studies agree that the doubling carbon dioxide concentration in the atmosphere from the pre-industrial level of 270 PPMV will probably increase global mean temperature by 1.5°C to 4.5°C but most likely by 2°C.

Since CO_2 is the biggest contributor to the green house effect and the first to be identified, experts measure the impact of trace gases in terms of the CO_2 equivalent. In 1980 the green house gases (methane, troposphere ozone, CFCs, extra water vapour resulting from green house warming of the oceans, and nitrous oxide) would have produced the amount of heating equal to extra 40 PPMV of CO_2. By 2030 their effect is expected to be the same as an extra 140 PPMV of CO_2. These gases therefore accelerate the warming of the earth. CO_2 concentrations alone may not double until 2080 but the addition of these gases would lead the same effect by 2025 some 50 years earlier.

Predicted temperatures rise due to concentrations of CO_2 and other green house gases by the year 2030 might be 3°C of which about one half would be caused by CO_2 itself.

Increasing CO_2 raises the temperature causing:

(i) More moisture to evaporate from land and ocean. The extra water vapour in turn traps more infrared radiation and causes the green house effect. The water green house feedback has been estimated to amplify the warming of the air by a factor of about 1.5 and the warming of earth's surface by a factor of 3.

(ii) Another well defined effect is the ice-snow albedo feedback. The green house warming will lead to more melting of ice or snow and so absorbs more solar radiation, thus increasing the warming. The ice-albedo feedback increases the global warming by 10-20 per cent but locally near sea margins and in polar oceans it can exceed the global warming by factors ranging from 2 to 4.

(iii) Increased moisture from warm oceans could alter cloud distribution and its characteristics. The nature of these cloud changes and their effect on the radioactive heating is not clear.

An analysis of surface temperature records during the past 100 years indicate that the earth's temperature rose by 0.5°C between 1880 and 1940, and had fallen by 2°C between 1940 and 1965 and had rapidly increased since 1965. These analyses suggest that global mean temperature has risen from 0.3°C to 0.7°C in the past 100 years. But a definite association between warming and green house effect requires further data and studies.

Since the global climate including rainfall, storms, wind patterns, ocean currents and sea levels ultimately related to global heat flows and temperature patterns, there is a general expectation that the earth's climate may be significantly modified in the next 50 years.

The reduced heat retention on account of ozone depletion may approximately equal that of global warming of CFCs.

There will be gradual increase in global mean temperature during the next century at the rate of 0.3°C $\pm$ 0.1°C per decade which is greater than that observed, over the last 10,000 years. The rise may not be steady because of many factors but may mean a 1°C rise by 2025 and 3°C rise by 2100. These temperature increases may be less if there are progressively increasing controls of CO_2 emissions.

The United Nations Inter governmental Panel on Climate Change (IPCC) projects a rise in average temperature to about 1°C–3.5°C by the year 2100.

Changes in Sea Level

The rise in global temperatures due to the effect of green house gases has caused the rise of sea level due to thermal expansion of sea water and melting of ice. In the past, climate has profoundly affected sea level. Geologists generally recognise that ice ages removed enough water from oceans to lower the sea by 100 metres below present levels. Although the glaciers that once covered much of northern hemisphere have retreated significantly, the world's remaining ice cover contains enough water to raise sea level by more than 75 metres.

It has been estimated that a global warming of 1.5 to 4.5°C would lead a sea level rise of 20 cm to 140 cm mainly through expansion of water. Climate warming could also melt the drifting ice in Arctic region and thaw the west Antarctic ice sheet. Melting the ice would substantially increase sea surface temperature and so shift major climatic zones 200 km or more northward. Recent studies conclude that a significant melting would lead a much larger sea level rise but that is not expected during the next century. There is a considerable debate over the likelihood of Antarctic ice sheet melting.

A sea level rise of about 1 or 2 metres would permanently submerge wetlands and lowlands, river deltas, mangrove forests and coral islands, accelerate coastal erosion, cause coastal flooding, threaten coastal structures and increase the salinity of coastal estuaries and aquifers. These would have far reaching environmental, economic and social implications for many centuries.

Without significant control over CO_2 emissions the global sea level may rise by about 6 cm per decade over the next century (Within a range of 3 cm-10 cm). Accordingly the global sea level may rise by 20 cm in 2030, and 65 cm by 2100, with significant regional variations. A global warming will constantly accelerate sea level rise and alter ocean circulation patterns and marine ecosystems. A 30-50 cm sea level rise which might occur by 2050 will threaten lowlands and coastal zones, while a one metre rise will make island countries uninhabitable, flood and contaminate fresh water and displace millions of people. A billion people may become environmental refugees by 1 or 2 metre rise. Maldive islands off the west coast of India and some Carribean nations would perish. Besides 1.88 per cent of West Bengal, 0.92 per cent of Gujarat and 4.6 per cent of Goa may be submerged. A large area of Bangladesh, China, Vietnam and Egypt may be submerged. Scientists warn that the east-coast slopes of India are gentler so they suffer more inundation. The stretch of west coast between 18° N and 10° N is safest on account of steep slope. South and north of these latitudes the inundation is more likely particularly in the east coast.

One estimate put the rise of sea level at 24 to 117 cm by the year 2050 and 56 to 345 cm by 2100 A.D.

Some scientists are sceptical about the rise of sea level due to melting of ice. According to Prof. Wingham from University College of London, who lead the international team of British, Dutch and US scientists carrying out research believes that sea level rise is most likely due to thermal expansion of water but not due to melting of the Antarctic ice sheet. Research from

Antarctica suggests that the idea of global warming due to melting of ice is a misconception. The findings show that over a period of five years no significant change in ice sheets thickness occured at all. He doubted the ability of glaciers to produce so much water.

But the scientists of University of Alaska are of the opinion that Alaskan glacier is melting and the front retreated by 12 km in the last sixteen years. This is due to warming up of cold regions and since the climate is changing. They believe that the average surface temperature has been gradually rising over the last century.

Climatic changes may also occur regionally with the rise of global temperatures. Comparisons of warm and cold regions, and computer calculations show that when the earth heats up most, the equatorial regions heat up least.

Such temperature changes would alter rainfall and wind patterns. Some parts of the earth would get wetter, some drier, reliable winds that bring monsoon rains may change course, in some regions storms may become more common. Climatic changes may prolong droughts in semi-arid tropics by temperature and decreasing rainfall in one or more seasons, worsening their already critical problems. Some scientists have already speculated that prolonged droughts in Africa may be an early manifestation of regional impact of global warming.

In the humid tropics rainfall may increase. Tropical storms may extend into regions where they are relatively common.

Changes Predicted due to Global Warning

The regional climatic changes will differ from global although there is uncertainty about these regional changes. The following changes may occur as a result of global warming:

The temperature rise in southern Europe and central North America will be higher than the global average. These regions may have reduced summer rainfall and soil moisture.

The predictions for tropics and southern hemisphere have greater range of uncertainty. The projected temperature and precipitation changes suggest that climate zones could shift several hundred kilometres towards poles in the next 50 years, a rate which is too rapid for flora and fauna patterns to adjust. Accordingly there may be major changes in the structure of natural ecosystems.

Many areas may have increased precipitation, soil moisture and water storage. On the other hand, water storage will go down in some areas and if they are already water starved such as Sahel Africa, the result could be catastrophic. The intensity of hurricanes may increase dramatically. Hurricane intensity is linked to the temperature of sea surface. If it warms up to predicted levels, the most intense hurricanes will be 40 to 50 per cent more severe than the most intense hurricanes of the past 50 years. Increased precipitation may be by 7 per cent more due to more evaporation of sea water.

One climatic model predicts that central India will have doubled precipitation. Mid western USA may have much drier summers than they have now. Some arid areas including California and Morocco will have drier winters. Drier soils cause even hot air. The impact could be even greater at the poles where mean temperature might rise to as much as 10°C.

Warming trend will change the temperature gradient between the cold polar regions and the warm tropics. As a consequence the traditional patterns of winds and ocean currents will shift.

Although computer models predict the total rainfall worldwide likely to increase slightly, the mid continental regions of USA, China, southern Canada are likely to become hotter and drier especially during the summer months. In India even more violent cyclones and floods may occur on the east coast.

So far there has been no clear evidence that warming trend has begun.

Effects on Territorial and Marine Ecosystems

Examining patterns of global vegetation and climatic changes in the distant past, leaves no doubt that a doubling of CO_2 concentration in the atmosphere could potentially have profound effects on global ecosystems.

In semi-arid regions trees are vulnerable to decrease in rainfall. In wet forests, they are vulnerable to insect pests and infestations influenced by temperature and precipitation. In the

Arctic Tundra, warming would go further north. The upper layer of Tundra peat would dryout increasing oxidation and decay of organic matter. More CO_2 is released and this would in turn enhance the warming, creating a positive feedback.

Food production and distribution would be greatly affected by climatic warming in general, increasing CO_2 concentrations in the atmosphere would have beneficial effects on crop yields. Laboratory experiments indicate that if there was no climatic change, doubling of CO_2 concentration would increase the growth and yield of maize, sorghum, and sugar cane by upto 10 per cent with a 10-50 per cent increases for wheat, rice and soyabean, depending upon the specific crop and growing conditions. Doubling the CO_2 concentration would reduce transpiration by 34 per cent, plants would use water efficiently. But if there is change in the climate, crops would be substantially affected. But crop impact analysis show that warm average temperatures decrease, both wheat and maize yields in the vital mid-latitude crop region of North America and Western Europe. Warmer and longer growing seasons induced by climatic changes may increase the rise of many insect pests. The increase of these pest populattions under the newly favourable conditions could increase crop losses.

The global warming produced by green house gases will arrive slowly. By the time the average temperatures have risen by one or two degrees agriculture should have largely adapted to the change. Different crops, different varieties and different farming techniques will all help to absorb the impact of change. The world agricultural trading system can also adapt to change as different countries export and import quantities of different commodities. However marginal food producing areas may have more difficulty in adapting. Their production is already very sensitive to climatic change, and the change of their agricultural technology is very slow.

Since CO_2 is responsible for about 50 per cent of green house effect, priority should be given to strategies that would freeze or reduce the rate of its emission including energy efficiency and modification of technologies that lead to reduction in the use of fossil fuels. The present preservation of forests especially the tropical forests is probably even more practical. Each of these measures can serve more than one environmental and economic purpose. Even if the worst scenario does not come about we would benefit in other respects by taking these measures now.

While global agricultural potential increase or decrease, there could be significant regional declines in productivity in Brazil. Peru, Central North America, Africa, South-East Asia and Central Asia. But there could be increased in cereal production in Northern Europe.

Policy measures and research could lessen the regional impacts so that global food production may be maintained. Forests may be adversely affected because the climate changes will occur in time spans comparable to their rotation periods so that there is little scope for adoption. In addition, the host-parasite relationship may change. These stresses may be greatest in some arid regions.

There could be dramatic disruptions of agriculture, water resources, fisheries, coastal activities and energy use. The unprecedented rapid change could accelerate the already high rate species extinction as plants and animals fail to adapt quickly enough. A temperature rise of 2°C in the tropics could reduce rice production by more than 10 per cent.

The nations most likely to reap benefits of warmer climate are Canada and Russia since much of their vast land area is too cold for large-scale crop cultivation. The computer models suggest that ecological zones shift northward. As mentioned already, the western desert belt may shift to the grain belt and the grain belt to Canada. There may be some gainers or loosers if the shift takes place.

It is possible that US corn belt will shrink in size because of higher temperatures and less rainfall. A study of scientists of Utah State University indicates that the cropping area for staples like corn and wheat in the US great plains could be reduced to one-third. Such declines could in theory offset by expanding agriculture in Canada and Siberia. As temperatures warm these regions which have poor soils, it would take centuries before they reach the productivity of current agricultural land.

In a world in which population is expanding rapidly, global warming is likely to cause the million in jeopardy. Crop yields are reduced at higher rates due to evaporation of soil moisture which is key to plant growth. Decline in monsoon rainfall in central northern plains of India is responsible for reducing soil moisture. Though the yields of some crops may increase, the nutritive quality of crops will go down because of decreased intake of nitrogen.

A British scientific study reveals that climate is getting warmer and the effects of global warming is noticeable in all flora and fauna of UK. The scientists are of the opinion that English oak will disappear from southern parts of England. Vineyards are being planted all over UK including colder regions in the north such as Yorkshire.

Frogs and birds have been laying eggs earlier than past. Scorpions, native of warmer climates are thriving in the southern country of Essex.

Apart from these, coral reefs in Andaman sea are loosing colour owing to extensive bleaching. Experts say this is due to global warming. Under bleaching conditions corals loose their growth and reproductive ability.

Decline in krill population in Antarctic waters which occured gradually over the past decade, matches a corresponding decline in winter sea ice development caused by warmer air temperature. Krill are small shrimp like crustaceans usually found in frigid waters of Antarctica. Krill found in sea around Antartica form a crucial food part of marine food chain. But the krill population is on its way to extinction.

Measures to Control Global Warming

Firstly, policies either for reducing emissions of green house gases or mitigating the adverse effects of climate as and when they occur or both should be formulated at the global level only. This is because green house emissions are readily dispersed in the atmosphere with the result that emissions from any point on the globe affect all points. In other words, emission of gases with equal global warming potential from different points cannot be distinguished in terms of their efforts. Further significant unilateral reductions in emissions by even largest emitting countries by themselves will not have major impact on building up of green house gases in the atmosphere.

An important aspect is to bear in mind while framing policies in reducing green house gases, is the major economic costs. This is because one cheaper carbon intensive fuel, principally coal would have to be substituted by costlier alternatives, such as nuclear power, or natural gas or renewables such as solar energy or wind power.

Secondly the growth of industries such as iron and steel which emit large quantities of CO_2 may be curtailed. Since different centres have different levels of responsibilities for causing global warming, their future rights to limit green house gases should also be differentiated. This may result in changing pattern of international trade, with countries having relative emissions of green house gases rights, gaining a comprehensive advantage over others with respect to green house gas intensive industries.

The third major policy dilemma is that while there are greater uncertainties in the timing and patterns of climatic change and its impacts, any changes in green house gas emissions may involve a major economic crisis. On the other hand, there are serious risks of catastrophic impacts, for example, major sea level rise and increase in desertification if nothing is done to curb emissions of these gases. Accordingly, the policy maker is faced with hard choices between a watch and see approach which may involve serious risks or do something now even in the absence of scientific certainty which can involve major economic costs.

Realising these problems and the need for cooperation, the United Nations adopted a two pronged strategy to the issue of global warming. Firstly, the United Nations Inter governmental Panel on Climatic Change (IPCC) was set up in 1988 under the chairmanship of a eminent meteorologist, Beri Bolen to reach international scientific consensus on different aspects of climatic change. Two major sets of reports of the IPCC (1990 and 1992) have been published indicating the areas in which there is scientific agreement and those in which there is un certainty. The international negotiating committee (INC) for climatic change was set up in 1990 to formulate a framework convention (or agreement) between countries to form the basis for schemes of international regulation of global warming. The INC had a number of sessions and some agreement containing a set of broad principles reached.

To sum up, climate change is clearly the most important global environmental problem yet to be faced by mankind. The potential adverse impact of global warming is serious and may affect key life support systems of the planet. On the other hand, approaches to the problem may have significant economic costs which could fall unequally across countries.

The crux of the problem of reaching a future international agreement to reduce green house gas emissions is the question of equity or fairness. One doctrine which has significant support although there are rival formulations as well, is that enunciated by Ya Sumasa Fujii. According to this doctrine, all human beings should have equal rights to emit green house gases irrespective of when or where they live.

Elements of the Fuji doctrine have echoed in all approaches of several developing countries at the INC. The principle would allow significantly greater allocation of green house gas emissions rights to developing countries such as China and India both on account of larger population as well as small per capita emissions relative to industrialised countries. One needs to understand, at the same time, that a sudden application of this principle globally may result in serious economic dislocation in a number of countries.

To sum up all the countries should view that global warming is a great environmental threat in future and thus, take all measures to reduce and finally act down completely the emission of green house gases. In this connection there should be an understanding between the rich developed countries and the poor developing countries. Industrialised countries have almost reached the saturation point since they have been emitting green house gases since the dawn of industrial age. But the developing countries, on account of low industrial development are emitting a low percentage of these gases.

The developed countries should stop the emission of these gases immediately by adopting pollution free technologies in their industrial and transport development, and they have been already going ahead with alternatives. But the developing countries with the existing pollution technology need more time to shift to non-pollution technology for two reasons. Firstly the per capita emission of these gases is very low when compared to developed countries. As such, need more time to go ahead with their industrial development with the existing technologies. If they suddenly stop emission of these gases, by stopping their industrial development, a large number of people will be unemployed. Therefore, they have to continue their industrial development with existing technologies for few more years until the pollution free technology is supplied at cheaper rates by developed countries since the developed countries are the chief polluters hitherto, by which the developing nations are also sufferers even though they are less polluters. It is their moral responsibility to provide the pollution free technology at lower costs, besides providing financial assistance to these countries. These conditions have been put by the developing countries in Earth Summit conference held in Rio de Jeneiro, Brazil, in 1992.

The developing countries also should develop pollution free technologies in their countries instead of depending too much on developed countries. In India many of the research institutes are engaged in this regard.

The developing countries should take drastic steps in arresting population growth. China has adopted zero population growth. India is lagging behind very much in this regard. Unless population growth is controlled, it is not possible to conserve resources and check pollution.

The methane emissions particularly from China and India have to be minimised by adopting new technologies particularly in cultivation of rice.

To minimise CO_2 emissions in developing countries, destruction of forests for shifting cultivation and for fuel should be avoided. Afforestation programmes should be undertaken. Wood, as a source of fuel should be substituted by alternate energy in these countries. Thermal power dependency should be minimised, pollution free vehicles should be used.

Change in outlook and lifestyle is needed in both developed and developing countries. Developed countries should consume less beef since Brazilian forests are destroyed to feed cattle for supplying beef to USA. America destroys 220 mil trees per year only for the manufacture of newsprint paper. Developed countries should reduce consumption of resources to check pollution. They must use less number of cars and consume less power. One should optimise satisfaction by minimising wants, instead of maximising output and over exploitation of resources for ever increasing wants. There is need to create environment awareness among people in both developed and developing countries on a war footing. There are no international boundaries so far for the spread of pollution. What one does in his country may have adverse affect in another country. Therefore one should think globally and act locally.

Global warming and its adverse impacts could be controlled with the mutual understanding and cooperation in all countries.

3

OZONE: THE SAVIOUR AND KILLER GAS

Ozone, being a life saver as well as killer is considered to be a gas of paradox in the atmosphere. Before we understand the various aspects of this gas, it is better to have a brief idea about the atmosphere covering our planet earth.

Earth as one of the planets of solar system was considered to be in a gaseous state in the beginning and it gradually reached the liquid state and then finally into solid state by cooling. When it reached the solid state, some of the lighter gases like hydrogen and helium were left out in space. The present gases found in the atmosphere were not there in the beginning. At present about 99 per cent of the atmosphere is composed of nitrogen and oxygen. Roughly 1/5th of atmosphere consists of oxygen. Other gases like carbondioxide (CO_2), hydrogen, etc, have a small percentage. The present state of atmospheric composition reached about from precambrian period according to geological time scale, i.e., from 600 million years ago.

The composition of atmosphere reached the present stage due to volcanic eruptions, hot springs, vegetation, animal life, weathering of rocks, etc. During precambrian period oxygen concentration in the atmosphere was 1/1000 of the present day level.

In the present atmosphere there are a few layers. The lowest layer is called troposphere and the next layer above this layer is called stratosphere. The thickness of troposphere varies and is not uniform all over the earth. It reduces as it goes further from equator. While it is 18 km. at the equator it is about 3 km. at the poles. In this atmospheric zone the temperature decreases with altitude. It is in this zone, clouds form, condensation takes place, and rain occurs.

In the next upper layer of the atmosphere i.e. stratosphere, the temperature increases upto 65.5°C in the first 32 km. and decreases to –73.50°C in the next 32 km. In this zone, cloud formation does not take place.

Ozone (O_3) whose molecule consists of three atoms of oxygen is present as a thin layer in both layers of the atmosphere. Though it differs slightly from oxygen in its composition, whose single molecule consists of two atoms, it is altogether different from oxygen in physical and chemical properties. Ozone is a pungent gas unlike oxygen which has no smell. It is present both in the troposphere and stratosphere. The most interesting difference between oxygen and ozone is, while oxygen is a life saving gas whereever it is present, ozone on the other hand is only a life saving gas when it is present in stratosphere only but it is a harmful gas when it is present in troposphere.

Origin of Stratosphere Ozone

Ozone is found in stratosphere at a height of 25 to 35 kms.

There are different views of scientists regarding the formation of ozone. The most widely accepted view is that ozone is formed from oxygen through chemical reaction by the action of ultraviolet (UV) solar radiation on oxygen molecules. Oxygen molecules absorb UV rays and as a result they split up into oxygen atoms (O+O). These oxygen atoms then react with unsplit oxygen molecules (O_2+O) to produce O_3. As such, stratosphere ozone is created in a photochemical process which begins with the breaking of oxygen molecule. It is produced mostly in tropics because of intense solar radiation. From there it rises to the stratosphere to form a layer. Ozone is also considered to be formed through electrical phenomenon due to lightening. The total amount of gas is estimated to weight 3 billion tons.

As mentioned earlier, stratospheric ozone is a life saviour for the entire biotic community i.e. human beings, plants and animals. By absorbing Ultraviolet rays of (UV-B and UV-C) of short wavelengths it protects human beings from skin cancer and eye diseases like cataract.

Depletion of Stratosphere Ozone

The depleting of ozone was discovered a few decades back causing large patches of ozone holes. A large hole was discovered in the Antarctic region in 1970. This occurence of ozone depletion phenomenon was discovered by scientists by converting ozone into oxygen due to presence of chlorofluorocarbons (CFCs) present in the atmosphere. The depletion of ozone is also found in north polar and mid-latitudes in lesser degree when compared to Antarctic region.

It was in 1960, that a British scientist James Lovelock detected the presence of CFCs in the atmosphere with the help of a device developed by him. To his surprise one of his devices on a ship sailing in the southern hemisphere found traces of CFCs far away from any major release point. CFCs once released into the atmosphere never disappear.

Soon after, Lovelock presented his data in early 1970, Sherwood Rowland along with a post doctoral fellow Mario Molena of Chemistry department of University of California began to investigate what would happen to CFCs. They quickly realised that they were studying a potentially devastating problem. In 1974 they reported CFCs being added to the atmosphere were steadily increasing. According to their theory that CFCs remain intact in atmosphere for 50 to 100 years slowly drifting 12 to 50 kms into the stratosphere, by 12 to 50 kms. above the earth's surface. There the Ozone depletion takes place when the Ultra violet radiation breaks the CFC molecules tight chemical bonds freeing chlorine atom (Cl) which in turn attacks an ozone molecule (O_3) to form oxygen molecule (O_2) and forming chlorine monoxide (ClO). The chlorine monoxide eventually split by an oxygen atom. creating oxygen molecule and freed chlorine beginning the cycle of destruction again. They calculated that every chlorine atom removes 100,000 molecules of ozone from the atmosphere.

Rowland and Molina's theory became controversial among atmospheric scientists, and public reaction to their research was muted. For many scientists, the Rowland and Molina's work remained as a theory without empirical evidence. Studies sponsored by National Academy of Sciences concluded that their predictions were overstated. The safety of CFCs faded as a public issue though it continued to be an area of concern to environmentalists.

However, in 1985 startling reports of a hole in the ozone layer over the south pole revived the issue. Researchers of British and Antarctic survey had noted the decrease of more than 40 per cent in ozone layer every September and October starting in 1977. As CFCs destroy ozone layer, Ultraviolet rays reach the earth in increasingly dangerous amounts.

Research studies by scientists from USA, Great Britain, France, Argentina and Chile were conducted to find out the ozone conditions in the Antarctic atmosphere. The most crucial information however was gathered by one of several experiments actually carried into the ozone layer by ER-2, a plane more closely resembling a rocket. During August and September 1987, the ER-2 made 12 trips into the atmosphere over Antarctica collecting and analysing samples of air from altitudes of upto 17 km. By October 1987, a chemical picture of Antarctica's atmosphere came out into focus as a result of these experiments. As the chlorine monoxide level rises steadily through August and September it reaches 500 times more to normal concentration at certain altitudes and ozone levels drop substantially. At the same time the ozone hole seems to fluctuate with the continent's atmospheric conditions.

There is already some evidence of damage to the ozone layer in northern hemisphere. A study released in March 1988 in which more than 100 scientists from US and United Nations agencies analysed ozone data gathered by satellite and ground based measurements found that atmospheric ozone in the range of 30 to 60° N latitude, a band that includes the most populated areas of USA, Canada, Europe, Soviet Union, China and Japan had decreased between 1.7 and 3 per cent over a period of 17 years (1969-86).

Scientists are now generally accept that man-made chemicals are the cause of the decrease. Because CFCs remain long in the atmosphere, the damage to ozone layer will continue for long.

Though depletion of ozone is a worldwide phenomenon found by scientists, but a large-scale depletion of ozone leading to a big hole in Antarctica is due to the special conditions of atmospheric circulation around Antarctica, according to scientific investigators. The size of ozone hole in Antarctica is about 20 sq. km. Most of the investigators believe that special conditions of atmospheric circulation around Antarctica in particular, the dynamics of the polar vertex, that circles the Antarctica during the polar night creates a region of very cold surfaces. Chemical reactions take place on the cold surfaces of frozen particles at high altitudes. These reactions according to current theory precondition the chemistry of air, so that when the sun returns into the spring, conditons are just right for the reactions involving chlorine that remove ozone so effectively from the stratosphere. Everyone now agrees that both dynamics and chemistry are involved and that the whole system may be sensitive to small changes in the environment.

Scientific assessment of ozone depletion in 1991 reveals that overall decrease of ozone is about 3 per cent per annum. Ozone depletion continues in all latitudes except over tropics.

As mentioned earlier the man-made chemicals, i.e., chlorofluorocarbons are the chief source for ozone depletion in the atmosphere. With one per cent increase in CFCs there is 2 per cent decrease in ozone. The CFCs were first discovered by an American scientist Dr. Thomas Mudley who demostrated the synthesis of a CFC as a colourless, odourless, non-toxic, corrosive, non-inflammable chemical. These chemicals were used first by Dupont and general motor companies in industries. These were first used in refrigerators, since 1950 they were being used as propellants in aerosal cans. The CFCs are widely used in a variety of manufacturing goods because of their stability, inertness, non-toxic and longer life. The CFCs are largely used in refrigeration, airconditioning in aerosal propellants and in foam rubber plastic industries. They are also used to clean computer chips and for making polythene. CFCs are synthetic chemicals marketed for the last 60 years under the proper bands such as freon and genetion. They do not react with other chemicals, stable and do not burn

CFCs are non-toxic, non-inflammable, and highly stable. They are ideal industrial chemicals used in airconditioners, refrigerators, foam rubber, aerosal sprayers and polythene containers used for making food packing. Use of CFC chemicals in the beginning was highly limited. In 1931, 500 tons of CFC_{11} and CFC_{12} were manufactured. But the picture changed since 1960 as multiple users and large number of people began to use them. The manufacturing increased by leaps and bounds from 1,69,000 tons in 1961 to present level. The production of CFCs was 25 lakh tons out of which 55 per cent was consumed by US and 35 per cent by Western Europe. The rest was consumed by all other countries. China consumes 1 per cent and India consumes less than 1 per cent. The per capita consumption of CFCs is 1700 grams in US and only 3 grams in India. This is changing fast and with industries using, CFCs, consumption is expected to grow at 15 per cent per annum and the per capita consumption of CFCs in India by 30 grams by the year 2005.

Out of total production of CFCs in the world, 1/3rd is consumed by refrigeration and airconditioning industries, 30 per cent is used in spray bottles and aerosal cans mostly perfumes and very little for medical uses. In bottles they act as propellants to expel the contents and in foam cans as the agents that create foam.

With the use of CFCs mostly in the industries cited above, the concentration of CFCs in the atmosphere has gone up. Out of CFC 11, 12, 14, 114, 115, which are used in industries. CFC_{11} and CFC_{12} are most important, and are largely produced. The annual emission of CFCs into atmosphere is estimated to be 0.4 tons. The concentration of CFC_{11} which was 150 parts per trillion by volume (PPTV) in 1977 increased to 226 PPTV in 1986 and it is expected to rise to 1100 PPTV by 2030. The concentration of CFC_{12} which was 226 PPTV in 1977, rose to 392 in 1986 and it is expected to rise to 1800 PPTV by 2030. According to another estimate, the concentration of CFC_{11} may rise from 1379 to 2897 by 2050 and that of CFC_{12} from 2359 to 3828 PPTV by 2050. On an average CFC_{11} is increasing by 4 per cent and CFC_{12} by 4.3 per cent per year.

Ninety per cent of CFCs are manufactured by developed countries which consume bulk of the product. In India four firms are manufacturing CFCs.

Because CFCs remain in the atmosphere for decades, the damage to the ozone would continue for long. The CFC_{11} lasts for 75 years, CFC_{12} lasts for 110 years and CFC_{13} lasts 90 years and halon 1301 for 110 years. The CFCs used for fire fighting are known as halons.

Because of their relatively low boiling points the CFCs rapidly enter into the atmospheric wind systems. Unlike several other chemicals, the transparent, insoluble and non-reactive CFCs cannot be removed from the atmosphere by scavaging processes such as photo dissociation, rain out and oxidation. This gives a residence time over 40-150 years. During this time they rise by randum diffusion from troposphere to altitudes 25-40 km (Stratosphere) where they destroy ozone by the process already mentioned.

Recent data from NASA's Upper Atmosphere Research Satellite (UARS) indicate that ozone depletion is also affected from other chemical substances. The results show that aerosols of sulphuric acid thrown up by volcanic eruptions such as those of Elchichon in Mexico and Mount Pinatubo in the Philippines could be possible for ozone destruction outside polar regions in recent past. Sulphuric acid aerosols not only provide a reaction site but also help in stratospheric ozone destruction in another way. Nitrogen oxide which is normally present in the stratosphere can protect ozone by combing with chlorine monoxide and turning into a harmless form. Researchers now suspect that a reaction catalysed by sulphuric acid aerosols immobolizes the nitrogen oxide, blocking its protective role. The data further show that aerosols are also weakening another component of ozone protection—a reaction between methane and chlorine, that might otherwise form destructive chlorine monoxide.

The data collected from high flying research planes reveal presence of chlorine monoxide rich air in the stratosphere as far south as the latitude of Cuba. The UARS data revealed another phenomenon. Blobs of chlorine monoxide rich air were found to move away from the pole for several weeks in January and hang over northern Europe from London to Moscow.

Even if the ozone destroying volcanic aerosols fade away in a few years, the threat will not lessen, for the atmospheric concentrations of man made chlorine compounds are increasing unabated. Even if the use of CFCs is completely stopped by 2000 AD, the abundance of atmospheric chlorine will continue to rise, to about 4.1 p.p.b (parts per billion) by the end of the twentieth century according to United Nations estimate. All that chlorine, the latest findings suggest will do more damage than was assumed. It may mean ozone loss will be our legacy for much of the twenty first century.

Other chlorine containing compounds such as carbon tetra chloride, methyl bromide, methyl chloride are also detrimental to the ozone layer.

Despite the scientists being able to provide scientific evidence to the cause of ozone layer depletion, mostly due to industrial chemicals, there are certain questions which are not answerable satisfactorily. If industrial chemicals are solely responsible for ozone depletion, why does it occur over the Antarctic, a polar region with no habitation. In fact, it should have occurred in tropical and sub-tropical regions. But the satellite studies clearly reveal that these regions are not affected. In this regard, Dr Sidharth, Director, B.M. Birla Science Centre, Hyderabad, expressed his view that solar winds could be responsible and there should be an effort to study their impact. Studies from the astronomical angle are sporadic and we have not used any satellite to collect data in detail.

Harmful effects of ozone

As mentioned already, ozone depletion causes harmful effects on human beings, plants and animals. The damage is caused by Ultraviolet rays, which are prevented to reach the earth's surface by absorbing them by ozone. Human beings are affected by the incidence of skin cancer, cataracts Dr George Hitchcock, President of Auckland Cancer Society in New Zealand recently said that most common form of cancer is skin cancer in New Zealand and it is on the rise. Every year 15000 new cases of skin cancer are discovered. The US environment protection agency (EPA) has shocking results of a recent study which shows that in the next 88 years, 131 million additional cancer cases are expected due to depletion of ozone resulting in Ultraviolet rays entering the earth. It must be recognised that for every 1 per cent depletion of ozone, the intensity of ultraviolet radiation increases by 2 per cent. The only grace for people with darker skins is that they have a lower incidence of skin cancer with the same intensity of ultraviolet radiation than people with lighter skins.

Studies of 200 plant species revealed that two-thirds oт them are affected. These include members of bean, pea, cabbage families. Sensitive crops to UVB radiation are soyabean, cowpea, water melon, etc. Crops like wheat, rice and barley are comparatively less sensitive to the radiation. Exposure to high intensity of ultraviolet radiation causes reduction in leaf size, stunted growth, poor seed quality, poor yields and increased susceptibility to weeds and pests. The height of the plant size is reduced, the stem growth is suppressed. Leaves are more affected since they are directly exposed. Sixty per cent of leaf size is reduced.

As a result of high incidence of ultraviolet radiation due to ozone depletion, the algae productivity in Antarctic waters reduced by 6-12 per cent. It also caused damage to various forms of aquatic larve and other organisms. The long term effects of such changes in Antarctic waters could be catastrophic. For example the productivity of phytophlanton could affect krills, the tiny shrimp like creatures that lie at the bottom of the Antarctic food chain. Krills would in turn affect fish, birds and marine animals including seals and whales that feed on them and finally human beings who greatly depend on sea food. With the description of the whole marine food chain, the ecosystem will collapse.

Thus, the ozone depletion causes damage to the entire biosphere.

Besides, UV radiation will also affect synthetic materials like paints, plastics and PVC, etc.

How to save ozone from depletion

Manually Adding Ozone

The most obvious solution would be to manually add ozone to the depleted areas of the stratosphere. Unfortunately this is not a feasible solution. From what we know from Antarctic stratospheric mechanics, the ozone levels worked out a balance even with depleted levels. Any ozone that we artificially add is most likely to be destroyed and would not help our problem in the long run.

Destroying CFCs from Atmosphere

The second alternative and futuristic approach could be to destroy potential damage causing CFCs from troposphere with high power lazers located on mountain sides. CFC molecules can absorb upto 30 infrared photons before they are destroyed, so a power blast would have to be maintained. It is thought however, that the infrared lazer beams would be shifted out of the needful frequency range before they got to the level of atmosphere where the CFCs were targetted to be destroyed.

Dumping Ethane or Propane into Atmosphere

Another probable plan could be to dump about 50,000 tons of ethane or propane into the Antarctic stratosphere each spring. The chemicals would transform active ozone depleting chlorine into non ozone depleting Hcl. The hydrocarbons (propane and ethane) unfortunately would decompose about in an year, so that this process would have to be repeated every year.

Using Alternatives to CFCs

Although we have a large number of CFCs like CFC_{11}, CFC_{12}, CFC_{14}, CFC_{114}, and CFC_{115}, the most important are CFC_{11}, and CFC_{12} which are produced in large amounts since they are used in several industries. In fact, these two are the most potential ozone depletion chemicals.

Today a vigorous effort is being made to prepare substitutes to CFCs which are equivalent to CFC_{12} in efficiency. The list of possible compounds favoured by the manufacturers are HEC-134a (CF_3CH_2F). Its properties are quite close to CFC_{12} and it will break down quickly in the atmosphere. Dupont and Imperial chemical industry (ICI) which produce 90 per cent of CFCs are vigorous by working out to find out substitutes. They are quite confident that it would be possible to phase out CFCs soon and replace these by HFCs (Hydrofluoro carbons) or HFAs (hydro fluoro alkalanes). Both these companies have established plants to make these compounds. The ICI has already adopted a trade name for its compound KLEA-134A and this will be the first of a family of alternative fluorocarbons (AFCs).

The race for CFCs replacement is on. All the manufacturing routes are bound to face some difficulties. There is a school of thought which argues that CFCs were introduced primarily to replace noxious fluids in refrigeration but today the damage caused by them outweighs their advantages. In view of this, it is to be thought how far will it be wise to concentrate exclusively on HFCs with their unknown and unpredictable hazards. The HFCs do have a small ozone depleting potential but their C-H bond makes them more easily hydrolised in the lower atmosphere. On the other hand the CFCs have long stratospheric lifetimes. The present approach aims to reduce amount of chlorine entering stratosphere.

The Indian Institute of Chemical Technology (IICT), Hyderabad, which is one of the leading institutes around the world is working in this direction. A substitute HFC-134 for ozone depleting substance i.e. CFC_{12} has been recently developed by IICT, Hyderabad and a plant has been setup in Hyderabad to start production from January 2000. Industries are rushing to find substitute chemicals and alternative technologies for survival. Even after developing substitute materials problems posed by technology shift would remain regarding:

(i) creating the necessary infrastructure to transfer it freely,

(ii) creating the necessary institutional infrastructure in developing countries to introduce the new technology, and

(iii) traising workers and technicians in developing countries in the use of alternate technology.

The National Chemical laboratory at Pune has identified to assess the properties of alternatives for airconditioning and refrigeration. In this process, IICT at Hyderabad, Indian Toxicology research centre in Lucknow, Central Mechanical Engineering Institute at Ranchi are involved.

International Conferences to phase out the cutting of CFCs production and use

The first global conference to counter the threat of CFCs in depleting ozone was held in Vienna in 1985. It called for protection of human health and environment. As early as 1977, the US decided to phase out the use of CFCs in aerosol cans. Canada, Norway, Sweden and many other countries brought a similar ban. But a proper protocol could only be drawn up in 1987 when 48 nations met in Montreal in Canada and signed an agreement and this is known as Montreal protocol. In this conference it was decided to reduce the consumption of CFCs by 50 per cent by 1999. It was also agreed to reduce production of five CFCs (11, 12, 113, 114, 115) and three Halons (Halon 1211, 1301 and 2402) to the level of 1986.

However, India, China and several other developing countries did not sign the Montreal protocol in view of unjust and discriminatory treatment to these countries. It was expected that developing countries would maintain the per capita consumption at the present level (10 mg in case of India). While developed countries were permitted to a per capita consumption of 500 grams till 1989 and subsequently it would be reduced to half by 1998. Besides, the developing countries were denied their request for free access to the technology for manufacture of substitutes.

However, in London conference in 1990 India and China were able to convince the international community of environmentalists of their just demands and signed the agreement on 20th June 1990. They had agreed to go along and phase out the CFCs by the end of the century after the developed countries set up a 250 million dollar fund to finance the developing countries to switch over to CFC substitutes.

Even after signing the protocol by developed countries, the worst is not over. Damage to earth's protective ozone will continue in the twenty-first century as the photoactive CFCs will continue to be used even though in decreased amounts. Besides, only some CFCs affecting stratospheric ozone are covered by the protocol and even those cannot be discontinued to the extent of 100 per cent.

Increasing scientific evidence has suggested that the Montreal protocol was inadequate in meeting the threat of ozone depletion facing mankind during the first week of March 1989. The countries decided to eliminate 100 per cent use of CFCs by 2000 A.D. Industries have been asked to find out alternatives. Dupont has already alternative materials for refrigeration, new refrigerators using FC22 which eliminates almost 90 per cent of ozone depletion are in the process of being introduced.

There are several problems relating to use of alternatives. The alternative manufacturing costs are five times more. They cost 150 billion dollars just for one country i.e. US.

London Conference (1990): Conclusions Drawn

The London conference held in 1990 came to the following conclusions as far as future of ozone depletion is concerned:

(i) There should be an international ozone protection authority which should collect data on every aspect of ozone and provide information to every country. It should intensify research on alternatives of CFCs which can be available without the patent rights to all countries. This would eliminate inequities in terms of available alternatives for developing countries.

(ii) The authority should prepare a long term global action plan which would protect ozone with the total ecosystem of the planet.

(iii) The authority should set up a system of taxing the CFC users according to the quantity consumed.

(iv) At the same time the developing countries should set up a third world agency which would deal with the critical issues of their development in relation to ozone depletion and other environmental problems. They should identify alternatives based on natural environmentally suitable materials in an integrated manner.

Developing countries for example can design insulation from jute materials. They should recover CFCs from old refrigerators and recycle them. They can also find alternatives for cleaning, processing for electronics by using natural chemicals from plants. Then again there are many uses of CFCs which are totally unnecessary for foam rubbers, for packaging and for furniture.

At the international conference, London, held in 1990, 54 countries had agreed to phase out the use of CFCs. US had agreed to stop using CFCs and carbon tetrachloride by 2000, methylchioroform by 2002, and HCFCs by 2030.

In 1992, the parties met at Copenhaegen had agreed to phase out all CFCs, carbon tetrachloride and methyl chloroform by 1996. Phasing out Halons still posed a problem since they have no substitutes.

An analysis of consumption of CFCs in developing countries shows that there are two categories. Those who have reached the consumption of 300 grams per head. Come under the first category of countries are those who shall never reach that level by the year 2000. The second category can be given a grace period of 20 to 30 years. This has been decided in London conference to give developing countries the year 2010 as the limit to phase out the use of CFCs.

A 15 year period is needed for developing countries in phasing out CFCs to reach 2 ppbv (parts per billion by volume) chlorine level in stratosphere, the goal of the Montreal protocol. The Indian industry would like the grace period to be long enough to synchronise with the development of substitutes with zero depletion level. An Indo US workshop on CFCs in New Delhi suggested to give 20 to 30 years grace period to be given to developing countries whose per capita consumption never reached 300 grams by 2000. Countries will first have to switch over to transitory substitutes and then have permanent ones. According to an estimate the switch over will cost India 2.5 billion dollars.

In our country the first CFC plant was set up in 1968, three decades after US. During the last few years three more plants came up. In India 70 per cent of CFC production goes to refrigeration and cooling, 5 per cent for foam blowing, 20 per cent for aerosols and the rest absorbed by others.

India's fear was if we ban CFCs in 10 years, we cannot make the present type of refrigerators and airconditioners. Apart from the household refrigerators, this technology is used by all the food processed industries. Unless the western patent holder of alternate technology gives it cheap, it is difficult to face the consequences.

Developing countries want to continue the use of CFCs even after 2010 in refrigerators and airconditioners. These countries stated that if funds are not available they are not legally liable to meet the agreed phase out schedules.

Protection Measures from Ultraviolet Radiation

Several countries took protective measures from UV radiation in areas where the ozone depletion is much. In Australia and New Zealand people wear thick cooling glasses for eyes and hats to cover head and ears to prevent from exposure to sun rays. In South Chile, children are not allowed to move in sun between 10 a.m. and 3.00 p.m. It is even advised to wear cotton clothes.

Ozone protection is a must at any cost to save the biosphere. Every country should give utmost importance to discover substitutes to ozone depletion chemicals.

Harmful Ozone

As mentioned earlier, ozone is also formed in the lower layer of atmosphere i.e. troposphere at 10 km height from ground level and it is quite harmful to human life unlike stratospheric ozone which is a life saviour of entire biotic community. In this zone, ozone is formed by burning firewood, coal, combustion of petrol and diesel oils and presence of nitrous oxide, methane gas emitted from cattle and rice fields. Besides these, others like gamma radiation used in food preservation plants, Ultraviolet lamps used for sterilisation, dermitological photo therapy components, high voltage equipments, photo copying machines cause ozone formation. The smoke emitted from jet planes is also a great contribution for its formation in this atmospheric layer. Its concentration seems to be high in large urban settlements where dense transportation network and more industries are present. The troposphere ozone partly comes from stratosphere also by descending into it. Its concentration at present varies from 0.04 parts per million (by volume (ppmv) to 0.07 (ppmv). If the concentration of troposphere ozone increases further, there is every likelihood of its causing certain diseases relating to lungs like forming holes in the lungs, asthma, chronic bronchitis, breathing trouble and diseases relating to heart and brain affecting nervous system, causing dullness, drowsiness, premature ageing, visual impairment, etc and stomach diseases like enlargement of spleen are caused. If the air in the cabin of high flying aeroplanes is not purified, the passengers and crew are subjected to these diseases.

Ozone's presence in the lower layers in smogs which are caused due to air pollution, mostly in urban areas is a great threat to human health.

There is need to develop such transport and industrial technologies in the troposphere not only to minimise the ozone concentration in the troposphere but also its deficiency in the stratosphere to save biotic community.

4

HAZARDOUS WASTE DISPOSAL: A GREAT THREAT TO ENVIRONMENT

What are Hazardous Wastes

Industries which contribute a lot for economic development of any country contribute large quantities of waste material, besides producing several finished products. Certain wastes are toxic and are very harmful and as such they are defined as hazardous. Hazardous wastes are defined as wastes that are injurious to human health or environment when not properly disposed off. The term hazardous has been used differently in different countries. In USA wastes are defined as hazardous if they significantly contribute to mortality or cause serious irreparable illness or pose a substantial present or potential hazard to human health or the environment when imporperly treated, stored or transported, disposed of or otherwise managed. In Germany (former FRG) the term 'special wastes' is used. The technical definition of such wastes is that 'because of their nature, composition, or quality, they are especially dangerous to health, air or water quality, are explosive, flammable, or could promote infections diseases, and therefore special requirements for their control are necessary. They are also generated by the process that converts harmless materials into toxic. It seems 90 per cent of these wastes are liquid or semi-solid. Some of these wastes are non-degradable and may persist in nature indefinitely.

The World Health Organisation (WHO) has defined hazardous waste as having "physical, chemical or biological characteristics, which require special handling and disposal procedures to avoid risk to health and/or other adverse environmental effects". One approach to the problem of adequately defining hazardous waste is to draw up a list of known wastes that present no significant short term handling or long term environmental hazards and then to define hazardous waste as any waste not listed. This exclusion list approach has been used in some countries for example UK. However, listings of hazardous waste are more widely used for regulating purposes either with or without accompanying criteria. This inclusive list approach is currently employed in Belgium, Denmark, France, Germany, the Netherlands, Sweden and the USA. The list comprises wastes from certain industries, wastes containing specific components, or specific waste streams identified in the processes from which they originate. The inclusive list offers a greater degree of certainty but suffers from the disadvantage that there may be some significant omissions.

The Cairo guidelines and principles for environmentally sound management of hazardous wastes defined hazardous wastes as "wastes other than radioactive wastes which, by reason of their chemical activity or toxic, explosive, corrosive or other characteristics causes danger or is likely to cause danger to health, or the environment, whether alone or when coming into contact with other wastes are legally defined as hazardous in the state in which they are generated or in which they are disposed of or through which they are transported."

Thus, there was no internationally accepted definition of hazardous waste. This has led to two implications. The first was that estimates of hazardous wastes generated by industries in different countries varied widely depending on the definition adopted. The second was that regulatory procedures were complicated especially when hazardous waste was carried within the same country from one region to another with different regulations or transported from one country to another. In 1988, the Organisation of Economic Cooperation and Development (OECD) identified a core list of wastes to be controlled and all others which are considered to be or are legally defined as hazardous wastes in member countries.

Hazardous Waste Production

Estimates of hazardous wastes generated in OECD countries vary from 300 to 800 million tons a year or even more. According to some estimates 88 per cent of these are generated in USA. Germany (former F.R.G.), France, UK and Italy are the main generating countries in Europe. In 1986, 3.7 million tons of hazardous wastes are generated in UK. In Hungary 3.5 million tons are generated annually. Data from other countries, especially developing countries are believed to generate substantial amounts. Brazil, India, S.Korea, and China, might top the list. A survey in 1985 indicated that 700 industries along the west coast of Malaysia were generating hazardous wastes. The composition of hazardous wastes varies from one industry to another industry and the composition of waste generated in different countries varies with different mixes of industry. For example, in USA, 71 per cent of the hazardous waste is produced by the chemical industry, 8 per cent by petroleum refining and metal industry and the rest by other industries. In Hungary on the other hand 66 per cent of hazard waste is produced in the mineral industry, 17 per cent by the chemical industry and the rest by other industries.

Sources and Types of Hazardous Wastes

Sources of hazardous wastes are innumerable and vary considerabley in nature. Industries top the list. Infectious chemical wastes, spent fuel from nuclear power plants, incinerator ash containing beachable trace metals and discarded lead batteries, all fall into hazardous wastes. Dismantling of nuclear weapons, abandoning of the stock piles of chemical and biological weapons generate lot of hazardous wastes.

The United Nations Economic Commission for Europe (UNECE) held at Geneva in 1997 identified 12 toxic pesticides and chemicals (Andrin, Chlordane, DDT, Dieldrin, Dioxins, Endrin, Furous, Heptachlor, Hexachloro benzene, Polycholorin biophenyls and Toxophene) as persistent organic pollutants (POPs) and proposed a ban or restricted their use. The ECE commission represented by more than 50 nations from eastern, western Europe, Central Asia, N.America, Australia and S. America. POPs toxic chemicals widely used in agriculture and industries pose environmental and health threats. Having ability to resist degradation these chemicals are persistent in the environment. POPs are semi-volatile and therefore able to travel long distances in the atmosphere, thus exerting a global effect. These are also transported in the environment in low concentrations by movement of fresh and marine waters. This resulted in their widespread distribution across the earth, including regions where there have never been used. Thus humans, and other organisms are exposed to POPs everywhere in many cases for an extended period of time.

The Basal Convention to which a number of countries are a party met in 1998 and finally categorised substances into categories A and B. Substances included under A category are those which are banned for free trade. These are metal wastes consisting alloys of lead, wastes containing or contaminated by lead and its compounds, Zinc waste not included in B, waste lead acid batteries whole or crushed, unsorted waste batteries excluding mixtures only of list B batteries, waste mineral oil unfit for their intended use, waste oils, hydrocarbons, and emulsions. The substances allowed for trade included under B category are clean uncontaminated lead scrap, zinc ash and residues unless they exhibit hazardous characteristics, zinc containing drosses, slag from zinc production, zinc scrap that is not dispensable, waste batteries not confirming to a specification excluding those made of lead.

Hazardous Waste Disposal and Environment

Hazardous waste disposal poses a great threat to the environment. The wastes are disposed of in the country in which they are generated by adopting certain methods or by transporting to other countries and dumping there.

Disposal methods adopted within the country

The traditional low cost methods of hazardous waste disposal are land fills, storage in surface impoundments and deep well injection.

Deep well injection is more common in USA than in Europe. Over 75 per cent of hazardous waste generated in OECD countries is disposed of on land. Wastes may be disposed of in bulk or stored in drums, barrels or tanks. Thousands of land fill sites and surface impediments used for dumping hazardous wastes have recently been found to be entirely unsatisfactory. Corrosive acids, persistent organics and toxic metals accumulated in these sites for decades as little thought was given to their environmental impacts. In USA about 76000 active land fills have been recorded and most of them unlined. The Comprehensive, Compensation and Liability Act—(The Super Fund Act) was passed in December 1980 to combat the situation. By 1987 the USA environment protection agency (USEPA) has placed 951 land fills, impoundments and other sites on its national priority list of sites needing urgent attention. Estimates indicate that the list may grow to 10,000 sites and the cost for cleaning them up may range from 23 billion, to as much as 100 billion dollars. In Denmark, 3200 abandoned sites have been recorded and the clean-up costs may reach 1 billion dollars. As many as 35,000 problem sites are found in former FRG of Germany, the expected costs are atleast 10 billion dollars. In Netherlands, there are 4000 abandoned sites, out of which 350 are demarcated for immediate remedial action.

There are several hundred cases of ground water contamination by chemicals leaking from unsatisfactory land fills and surface impoundments. In USA more than 2000 substances including 175 organic chemicals have been identified in ground water. Several serious incidents have been reported. Waste stored in unlined canals and ponds near Denver and Colorado contaminated ground over an area of about 40 sq. km. At Perham, Minnesota, some people were poisoned by arsenic that had contaminated a well.

Other unsatisfactory dumping has exposed people directly to hazardous chemicals. At Love Canal, near Niagara falls, USA, homes were built on a former dump containing pesticides, chemicals used in making plastics and sledge from the bottom of stills. The dump was sealed with clay and sold to the local community. Rain water percolating into the ground leached the chemicals into a sledge that contaminated the buildings. Hundreds of families had to be evacuated from the site and the clean up costs reached millions of dollars. Researchers assessing potential health problems in children living near the site found selzcres, learning problems, hyperactivity, eye irritation, skin rashes and stomach pain all were prevalent among the population. At Lakkerkerk near Rotterdam in the Netherlands, drums of paint solvents (aromotic hydrocarbons) were included in rubble used to reclaim land. Houses were built on it. Again hundreds of families had to be evacuated and corrective measures were taken. These and other incidents that hit the headlines are only a few of those that have actually occured and many more are likely to have gone unreported. Methyl mercury discharged from a chemical factory into the sea or produced in the sea from inorganic mercury discharges by marine organisms, contaminated the fish eaten by local people at the town of Minamata on Kyushu island, Japan. As a result of this and a similar incident in Niigata on the east coast of Honshu, nearly two thousand people suffered neurological disorders; about four hundred people died. Although dumping of waste at sea is controlled under international and regional conventions, several countries are still using this route for the disposal of hazardous waste. In UK, for example, marine disposal is the second most used route after land fill for hazardous waste disposal. In 1986, about 5,79,000 tons of waste was dumped at sea, half of it would be classed as hazardous. Following the second international conference on the protection of North sea held in London in November 1987, it was agreed that dumping of industrial wastes in the North sea should be phased out by 31st December 1989. It was also agreed to cut the use of marine incineration by 65 per cent by 1st January, 1991 and to phase out altogether by 31st December, 1994.

Underground strorage of hazardous waste is practiced on a limited scale in some countries for example, Canada, USA and Netherlands. The best known is Herfa-Neurode facility in former F.R.G. which was operated in 1972. About 2,72,000 tons of hazardous waste have been placed about 700 metres down in an abandoned part of a potash mine: annual storage is now running at 35,000 to 40,000 tons.

Most countries still rely on land disposal methods for their hazardous wastes. Several technologies are available for handling them. Physical, chemical and biological methods can be used to reduce the bulk or toxicity of the waste. Of all the treatment technologies available, properly designed incineration systems can provide the highest overall degree of destruction and control for the broadest range of hazardous wastes. The best known system is Kommunekemi facility in Denmark. Established in 1973, in the town of Nyborg, it destroys over 90 per cent of Denmark's hazardous waste. At the same time it recovers heat from incinerators to supply Nyborg's residents with 35 per cent of their heating needs. In Bavaria of Germany (former FRG) integrated treatment facilities equipped with incinerators, inorganic chemical treatment plants and secure land fills form the technological backbone of hazardous waste management. Similar incineration systems have recently been established in Finland, Sweden and South Korea.

Ideally incineration should produce carbondioxide (CO_2), water vapour and inert ash. In reality incineration involves extremely complex reactions which are aggravated by complex and fluctuating nature of waste itself. The USEPA conducted a survey of the composition of hazardous waste for incineration. 237 different constituents were identified in one or more of 413 hazardous waste streams reviewed in the survey. Incineration devices are designed to optimize the chances that these reactions will be completed, but they never completely attain the ideal. Small quantities of their products may be formed depending on combustion conditions and the chemical composition of the waste. These products, along with potentially untreated components of the waste constitute the incineration emissions. Emissions from incinerators include hydrogen chloride, nitrogen oxides, sulphur dioxide, trace metals and trace amounts of dioxin and furan. Studies have shown much emissions appear to pose little increased risk to human health, but more detailed studies are needed.

Incineration at sea has taken place in specially designed ships, the costs are one-third to one-fourth of the costs of land based incineration, since emissions are not as tightly, controlled. However, there is now a trend to limit marine incineration or ban it altogether. Rising costs, scarce treatment capacity and public opposition to siting new facilities plague hazardous waste programmes virtually everywhere, despite its demonstrated destruction capabilities and the apparently incremental risks of emissions. At Denmark's Kommunekeni facility, for example, the volume of incoming waste has mounted by about 17 per cent a year for the last few years. The incinerators are now operating at full capacity, a new one is needed. But local opposition may prevent it from being built in Nyborg and it may have to be constructed elsewhere which will delay bringing it on line. Similar public opposition has also been encountered in USA.

In spite of high cost, and the stringent regulations, incineration is a popular method for hazardous waste disposal. Resource conservation and regulations act (RCRA) 1976, regulations require a destruction and removal efficiency (DRE) of 99.99 per cent hazardous waste incinerators. According to Toxic Substance Control Act (TSCA) of 1979 regulations, wastes with Polychlorinated Biphenyl (PCB) content of more than 50 ppm require a DRE of 99.9999 per cent. To achieve these stringent standards many innovative thermal destruction technologies have been investigated. Circulating fludized bed incineration, oxygen enriched incineration, molten salt technology, infrared system, plasma arc technology and wet air oxidation are some of them.

Plutonium recovered from dismantling of nuclear weapons and from the processing of spent nuclear fuel can be burnt in nuclear power reactors as Mox (mixed oxides of plutonium and uranium) fuel. Another method used is Vitrification-Immobolization in glass or ceramics for underground storage. The thick blanket of clay and mud that covers the ocean basins are considered to be effective in storing the nuclear wastes. Sea floor clay particles bind radio nucleides, immobolizing them.

Base catalyzed Dechlorination (BCD) is a technology for PCB removal. It puts into reverse the process by which pesticides and organic chlorides are manufactured, turning them into a biphenyl and salt. The biphenyl can be used as a fungicide for wrapping paper. It seems ten BCD plants constructed would be far more economical than one high temperature incinerator.

Hazardous Waste Disposal in Foreign Countries

Transport of hazardous waste from one country to another either for disposal and/or treatment takes place. Some of the developed countries have transported their hazardous wastes to some developing countries as well as developed countries. More than tenth of waste generated in countries is transported to other countries. This involved some 20 to 30 thousand cross events a year. This transport is due to absence of facilities in countries of origin and also due to high costs of disposal in these countries.

Some of the developing countries with low economic status and some developed countries with treatment facilities enter into business contract with exporting countries to earn money.

With regard to export of hazardous wastes into developing countries unnamed Italian and Swiss firms signed a deal with Somalia. According to contract Switzerland shall build and operate a facility for incinerating upto half a million tons of hospital and industrial wastes per year for a period of 20 years. A Swiss firm built two ocean incinerators off the Somalia coast. Somalia was to receive 80 million dollars for taking these wastes. Disposal of these wastes in Europe would cost 2500 dollars per ton, but barely 1 per cent of that to shop for the African dump.

A UK company got a 10 year licence to dump 10 million drums of wastes from Europe. In the Annobonian island in Equatorial Guinea for a fee of 1.6 million dollars. Benin entered into a contract to receive 5 million tons of waste per year receiving 2.5 dollars per ton from the US and Europe. In another agreement with France it was to receive 18 million dollars for accepting hazard wastes. It also received large quantities of radioactive waste from erstwhile USSR. Gabon an African country had agreed to take Uranium tailing waste from Colarado mines.

Illegal dumping of hazardous wastes of developed countries into some developing countries without the knowledge of these countries had also taken place. There are several instances which have occured. Recent examples of dumping hazardous wastes in developing countries are 15,000 tons of industrial incinerator ash from Philadelphia, USA on the Guinea island of Kassa, upto 4,000 tons of chemical wastes from Italy in the port Koko in Nigeria, about 2,500 tons of wastes from Italy in Lebanon and hazardous wastes from USA, Japan, former FRG and Singapore were dumped at Bangkok in Thailand.

As mentioned already, hazardous wastes are also transported between developed countries for treatment and safe disposal when these facilities are not available. About 80 per cent of the international trade is between the developed nations only.

UK imports of hazardous wastes for treatment, for example, rose from 5,000 tons in 1981 to about 53,000 tons in 1986-87. The largest exporters to UK were Netherlands, Ireland, and

Belgium. Lesser quantities arrived from Portugal, Canada, USA, Spain, Sweden, Italy, Australia, Singapore, Hongkong and Switzerland. France imported electrical equipment contaminated with PCB from Australia which lacks the technology to treat these wastes. It has been reported that waste importing industry is growing in France. Japan shipped the plutonium extracted from the spent nuclear fuel from France. It was to fuel its fast breeder reactor and conventional nuclear power stations. This is a part of the strategy for closing the nuclear fuel cycle by recycling the spent fuel. There were protests from all over the world against this shipping.

India's Position in Respect of Hazardous Wastes

Like many other developing countries, India also receives lot of hazardous waste material consisting mostly lead ash, battery scrap, zinc ash, waste oil and even old ships with toxic polychlorinated biphenyls (PCBs) and asbestos. The country imports 70,000 metric tons of zinc waste and 50,000 metric tons of lead waste through its major and minor ports. Even though the government says that the situation is under control, the ground reality is quite different. We are becoming the waste destination of the world a senior customs official commented. Due to skeletal staff and a lengthy coastline it is becoming difficult to protect India from waste imports. In spite of court order the import of hazardous waste is taking place.

Cheap labour, poor environmental standards, a sieve like import regime and growing market for cheap raw material are responsible for India becoming a heaven for dumping waste. Customs authorities admit that once a consignment lands on Indian shores, the issue of legality becomes a theoretical one. There is no provision under the Indian law for the consignment to be returned to the sender.

Delhi one of the over 1000 warehousing locations has currently 108 containers of hazardous wastes lying unclaimed. Alang in Gujarat has become the world's largest yard for breaking old imported ships. India is one of the 118 signatures to the Basal convention.

Environmentalists and the centre for Indian trade unions held protests against the decision to allow export of toxic laden ships of the US. The protesters alleged that the ships were contaminated with asbestos, bad and pollutants such as polychlorinated biphenyls (PCBs). Environmentalists say that both Basal convention and the Indian hazard rules classify asbestos or wastes containing PCBs as hazardous. Export of such wastes from OECD to Non OECD countries is illegal. The US is dumping hazardous wastes in South Asia to escape environmental and worker safety liabilities. US and Germany are the only countries in OECD whose hazardous wastes are known to have entered India after supreme court injunction on May 5, 1997 banning the import of hazard wastes.

Lead acid batteries export to India by US was around 467.1 tons in 1994-95. Smelters buy these batteries at a price of about Rs. 9 per kg. while the recycled lead sells at around Rs. 28-50 per kg. The main smelting centres are Kanpur, Calcutta, Hyderabad, Ghaziabad, Mumbai and Delhi. Leading exporters of these wastes are US, Canada, Germany and Switzerland. These junk batteries purchasing for recycling creates heavy lead pollution. This is the most common toxic waste that finds its way into India.

Recently the Swedish companies were suspected of allegedly dumping toxic industrial waste in India. Because of the Swedish laws of commercial secrecy, names of the companies involved were kept secret. Under Swedish law no company is permitted to export such materials to India. Most of these waste materials exported were metals for recovery. According to Green peace a voluntary environment organisation between October 1994 and March 1996, about 46,000 tons of zinc wastes and 21,000 tons of lead wastes were exported to India. The government claims only five companies are allowed to import wastes into this country but Green peace reveals about 151 companies import hazardous wastes into India.

On an average about 2000 tons of toxic substances are being dumped in our land every day by various industrial units without the requisite safe disposal facilities. Recently the supreme court banned the import of hazardous wastes for use in India. about 9000 industrial units will be affected as they are using the imported toxic wastes.

Measures to Control Hazardous Waste Disposal

Several countries had adopted legal measures by making laws. In the US the Resources Conservation and Recovery Act (RCRA) passed in 1976, was the first major hazardous waste regulatory document to address waste generation and disposal. It was revised in 1980 and 1984. The 1984 amendments known as Hazardous and Solid Waste Amendments (HWSA) significantly expanded the scope of RCRA changing many practices and policies. In 1979, the Toxic Substance Control Act (TSCA) prohibited the manufacture of polychlorinated biphenyls (PCBs). This established regulations for proper treatment and disposal of these wastes.

European parliament passed a resolution condemning all significant exports potentially hazardous wastes into developing countries. The US congress also prohibited the export of waste classified as hazardous in US unless special permit is granted.

Organisation of Economic Cooperation and Development (OECD) and United Nations Environmental Programme (UNEP) are trying to monitor and control the hazardous waste export to prevent developing countries becoming repositories of hazardous wastes. Existing OECD and European community requirements are such that transfrontier movements of hazardous wastes should be stopped unless an appropriate official of the importing country has provided written consent for the import and unless the wastes are directed towards an adequate disposal facility.

The organisation of African unity adopted a resolution in May 1998, condemning the use of African territory as a dumping ground. It called for a ban on the import of wastes into the continent. Several African countries and other parts of the Third World started formulating regulations of tightening up of existing ones to ban or restrict imports of hazardous wastes.

As mentioned already the Basal convention to which a number of countries are members, banned certain toxic hazardous waste items under category A for trade and allowed items under B category for trade.

To put an end to the import of toxic waste into India Central Board of Excise and Customs has directed the customs commissioners to test *prima facie* the hazardous imports. Such imports should be cleared only if the laboratory test reports classify them as non toxic. In case it is hazardous, clearance should be taken from the Ministry of Environment. This is an interim order passed by Delhi High Court banning the import of all toxic wastes.

Besides legal measures, technological and other remedial measures have to be adopted to prevent dangerous consequences of hazardous waste disposal. Thus, prevention is better than cure. Industries can avoid the costs and risks of treating, storing, transporting and disposing by reducing or stopping its production. Waste prevention or reduction is the best way to protect environment. If less waste is generated, it is cheaper to manage and there will be fewer failures in handling it.

Waste can be minimized by changing manufacturing processes, separating and concentrating waste, and reusing and recycling it. Other options include using different raw materials and replacing hazardous products with safe substitutes. Of course waste minimization means different things to different industries depending upon the particular waste streams being generated, their quantity and physical form.

Numerous case studies demonstrating the feasibility and cost effectiveness of waste minimization are available. One company has safely used pumice (a volcanic rock) instead of 20 tons of hazardous chemicals per year to scrub flexible metal circuits. Another has developed a process that uses ultraviolet light instead of hazard solvents to dry and set coloured paints. A third company has released a new line of non-hazardous bio-degradable industrial cleaners. There are many other examples. But despite such signs of a shift towards waste minimization, very little of the potential gains have so far been achieved. The USEPA estimates that existing technologies could reduce the total amount of hazardous wastes generated in US by 15-30 per cent by the year 2000.

Recycling and reuse of waste has been practiced in some countries for decades, for economic reasons. Perhaps the best known examples are the use of scrap metals and the reuse of glass bottles for soft drinks, etc. Recycling is now receiving increased attention in developed countries. Only about 4-5 per cent of hazardous wastes are being recycled in some developed countries, but there is a great potential for recovering several materials such as solvents and metals including chromium, mercury and copper. It has been estimated that upto 80 per cent of waste solvents and 50 per cent of metals in liquid waste streams in USA can be recovered by existing technologies. Japan seems to be leading among the major industrial countries in recycling and reusing its industrial waste, largely due to a unique cooperative relationship between industry and government. In Japan, USA and Western Europe waste exchanges operate on the simple promise that one industry's waste can be another's raw material have succeeded to varying degrees in promoting the recycling and reuse of industrial waste. Most serve as information clearing houses, publishing catalogues of waste available and waste wanted lists to inform industries of trading opportunities. A successful trade benefits both buyer and seller, the buyer reduces its raw material costs, the seller its treatment and disposal costs.

There are several technological approaches to deal with hazardous wastes generated by industries. Reduction in hazardous wastes could be a achieved through more vigorous research and development in waste minimization technologies and recycling, technical and financial support to encourage investments in them, and in some cases a tax on waste generated.

Few developing countries have established the basic foundation of a hazardous waste management system. Most have no regulations, no trained manpower, and no facilities capable of adequately treating and disposing of hazardous wastes. An active exchange of information and experience between developed and developing countries could do much to advance the latter's capabilities to deal with such wastes. Special emphasis should be made on strategies on waste minimization, recycling and reuse which would lead to vast economic and environmental gains. The general obligations in the draft global convention on the trans boundary movement of hazardous wastes and several of its articles dealing with exchange of information, technical assistance and control measures for the movement and disposal of hazardous waste give a much greater impetus to minimizing the production of hazardous wastes and the risk of dealing with them.

5

ACID RAIN: AN ENVIRONMENTAL HAZARD

Rain becomes slightly acidic with the carbon dioxide of the atmosphere but is not harmful to human beings, plants and animals. However, it reacts with certain rocks like limestone which consists mostly of calcium carbonate and decomposes. Even then it is not called acid rain. The degree of acidity is expressed in the scale number known as PH (which stands for the potential of hydrogen and is a measure of activity of hydrogen). It ranges from 1 to 14. A change of one unit represents a ten fold change and a solution say at PH_3 is ten times more acidic as one at PH_4. Pure water with a PH of 7 is neutral. Solutions with lower PH are acidic and those with higher PH are alkaline. Water acid mixtures fall on earth as snow, fog, hail, snow. Rain with PH value less than 5 is considered as acid rain.

Sulphur dioxide Contents

Scientists became aware of acid rain more than a century ago. The term acid rain was first discovered by a chemist Robert Angus Smith in 1852. The chemist while studying the quality of air around his home town Manchester, found local rainfall to be acidic. He suspected a connection between the occurrence of sulphur dioxide (SO_2) emissions when coal was burnt by local factories. He even noticed that the acid in the air was causing textiles to loose colour and metals to corrode. The air over Manchester was partly cleared by building tall chimneys at power stations and factories which release pollutants high in the air.

These chimneys improved the air quality by dispersing the pollutants, but they were only blown across into neighbouring countries.

The seriousness of the problem was not realised till Swente Oden, a soil scientist who noticed the severe acidification of Swedish water bodies and elimination of fish and other aquatic organisms. He found water extremely acidic and traced the cause to the emissions from UK and eastern Europe to Scandinavia. About 25 per cent of Sweden's water was acidified. Injected into high layers of the atmosphere, the pollutant gases are transported to surrounding countries depending upon direction, time and speed of the wind.

Acid rain consists mainly oxides of sulphur and nitrogen. Sulphur dioxide which is one of the chief culprits reaches the atmosphere through many routes. Sea spray, rotten vegetation and in some places volcanoes are important natural sources. All over the world perhaps half of the sulphur in the air comes from such sources and half from burning fossil fuels. Once the sulphur dioxide reaches the atmosphere it reacts with moisture in the air. First it is oxidised taking up atoms of oxygen to become sulphur dioxide and this combines with hydrogen to form sulphuric acid. Not all sulphur dioxide becomes sulphuric acid. A large part of SO_2 can travel in stable air with little signs of conversion to acid. In Europe most of the SO_2 falls to the ground unconverted and 85 per cent is added to atmosphere by burning fuel.

Nitric Acid Contents

Nitric acid is another chemical substance present in the acid rain. Nitrogen oxides are given off from power stations and car exhausts and include both nitrogen dioxide and nitric oxide which in itself oxidises and changes into nitric acid. Commonly sulphuric acid (7 per cent) and nitric acids (30 per cent) dominate. Ammonia from livestock sources and agricultural fields, hydrogen chloride from sea spray, volatile organic compounds (alkalines, polychlorinated aldehydes, esters and hydrocarbons) also add up to the acid rain. The photo oxidant ozone plays a catalytic role in the atmospheric transformations.

Stages in Formation of Acid Rain

A simple model of acid rain generation envisages six stages of reactions:

1. Emulsion of acid gases (SO_2 and NO_2) from natural and man-made sources at the ground level.
2. Some oxides fall back directly to ground level around the point of emission with delay as gases and aerosols (dry deposition).
3. Formation of ozone, the photo-oxidant by sunlight.
4. Interaction of ozone with SO_2 nitrogen oxides to produce H_2SO_3 (sulphuric acid) and HNO_3 (nitric acid) by oxidation.
5. Dissolution of sulphur, nitrogen oxides, photooxidants and other gases, including ammonia in cloud, or in rain water to produce acids. In coastal regions sulphates and chlorides from sea spray also add up.
6. Setting down of acid containing snow and rain at a distance far from the emission sources (wet deposition).

These sulphur and nitrogen compounds emitted by burning fossil fuels (coal; petroleum) can be blown thousands of kilometres to cause rain in countries far from the points of origin. These pollutants emitted from sources in the industrial heart land of mid-western US and Central Canada regularly fall on the comparatively less polluted areas of S.E. US and Canada.

Coal and oil burning release sulphur with which sulphur oxides are formed. Nitrogen oxides were also added forming nitric acid. Nitrogeneous fertilizers, fired forests and grasslands contribute nitrogen oxides. Industries releasing hydrogen chloride is converted into hydrochloric acid.

Some 150 mil tons of sulphur oxides and 40 mil tons of nitrogen oxides are currently estimated to escape into atmosphere by human action. A singly nickle-copper smelting plant located in Ontario, Canada emits 6,50,000 tons of sulphur oxides each year. Two–thirds of acid rain problem is attributed to sulphuric acid, nearly a third to nitric acid and the remainder due to hydrochloric coal and oil are the root cause for acid rains, particularly in India coal burning thermal plants and steel industry are greatly responsible. Already a low PH level is observed in Delhi and Chembur in Maharashtra.

India releases maximum pollutants into atmosphere after China among the developing nations. The sulphur emissions are expected to rise from 4,400 kilo tons in 1990 to 6,500 kilo tons in 2000, 10,900 in 2010, 18,500 in 2020. Low PH levels have been reported in Delhi, UP, Maharashtra, MP, Tamil Nadu and Andaman islands.

The use of large amount of diesel is causing a high amount of sulphur, nitrogen emissions in metro cities of India. Indian diesel has a sulphur content of 0.5 per cent by weight. Even this is far higher than sulphur levels in diesel used in countries like Sweden (0.001 per cent). The only good news for India is that chances of acid rain occuring are unlikely. This is because of tropical climatic conditions and predominantly alkaline rich soils of the country having a neutralising effect on the pollutants. As such dust particles in the country are alkaline in nature, acid rain causing gases such as SO_2 and nitrogen oxides get neutralised.

But if fossil fuel emissions go unchecked there is a possibility of SO_2 emissions increasing four–fold in India by 2020. Fortunately the sulphur content in coal in India is less when compared to China. In spite of that, there is a possibility of 315 per cent increase in sulphur dioxide emissions by 2020 as per the indication of IIT scientists.

Adverse Impact of Acid Rain

There is so much of adverse impact of acid rain on vegetation, birds, animals and human beings.

On Vegetation and Crops

Acid rain attacks bacteria in the soil which play an important role in reproductivity of plants. It also prevents essential nutrients from being used by plants. The photosynthesis process which produces food and releases oxygen into the atmosphere, can be hindered. This takes place when the protective layer of wax on leaves or needles is corroded by the acidic deposition. Besides, the ability of plants to resist pest attacks can be impeded. Acid rain quickly erodes the top soil ridding the earth nutrients needed by plants. Sulphur dioxide directly damages leaves and blocks stomata on leaves.

Microbial diversity in soils declined heavily and their activities were affected with chain effects on terrestrial plants.

Percolation of acid rain to the water table and subsoil water increases the acidity of ground water. This leaches the nutrients in it which are necessary for plant growth. It also increases the solubility of aluminium, mercury, lead and other metals which are toxic to growth. The overall result is damage to the foliage stunted root growth, increased susceptibility to disease and abortive germination. Plants are more sensitive to this kind of attack than animals.

Agricultural crops are also affected by acid rains. The yields of crops are lowered. In India high concentrations of sulphur dioxide around thermal power stations pose danger to the crops. C.K. Varshney, Professor of Environmental Sciences, Jawaharlal Nehru University has studied 26 crops including cereals, pulses, vegetables, and fruits around thermal power plants in India to guage the impact of SO_2 emissions on the yield and biomass. His study revealed that high level of SO_2 can reduce the leaf area by upto 40 per cent and shoot length by 14 per cent. There is a decline in wheat production near thermal power plants due to SO_2 emissions.

Water Bodies and Aquatic Life

Rivers, lakes and coastal waters in several countries of the world have been contaminated by the pollutants of acid rain. The mixture of aluminium and acid in lakes and rivers have profound effect on fresh water ecology. Fish have disappeared from thousands of lakes of Scandinavia since the 1950s. The deaths are due to aluminium poisoning.

Lakes, ponds and rivers became acidic and were unable to support biodiversity. Thousands of water bodies became barren without life. Fish death occurs primarily due to loss of ions of sodium and chloride and causes depolarisation of nerve and muscle cells which eventually cause death.

The species of green algae are eliminated by the increasing acidity. More acid tolerant species grow. In certain lakes a submerged aquatic moss and attached algae spread rapidly at the expense of acid sensitive species, bacteria in tolerant of acid stress are also eliminated.

Swedish workers have observed thick fundamental mats over the sediments in some acidified lakes. The sediment water interaction is greatly reduced by the growth of fungal mats. This hinders nutrient release from the sediments and the lake waters become progressively nutrient deficient.

The presence of aluminium in acid rain is highly toxic to many organisms.

Effect on Birds

Among birds the process of egg shell formation is disturbed. Calcium deficiency is known to hinder successful hatching of bird species. The ratio of calcium and aluminium is an important factor.

Effect on Human Beings

Acid rain water liberates mercury from soil which hinders brain development during the foetal stage. Fish eating birds and humans acquire mercury by eating fish with high levels of metal in them. The fish in turn ingest microorganisms which consume mercury released by acid rains in the water.

Acid rain also releases aluminium and cadmium. Cadmium can cause kidney disorders, besides accumulating in the outer layer of the kidney causing wounds. Aluminium on the other hand causes problems for kidney patients in dialysis (the process of purifying blood when the kidneys do not function properly). It enters the blood stream directly without first having passed the body's normal protective barriers. This may cause skeletal and brain damage. It may also cause Alzheimer's and Parkinson's diseases.

Nitrogen oxides besides producing nitric acid also react with other impurities in the air to make a mixture which has been called photochemical smog. This penetrates the lungs and interferes with the passage of oxygen into the blood.

If the water is drawn from ponds or lakes the aluminium leaching is likely to pose danger to kidneys. In Sweden, acid rain leached copper from water pipes contaminating drinking water and affected children's health.

Effect on Structures

The damage to bronze statues, limestone structures and sculptures are equally striking. The pitted roof of the famous 16th century Chapal in Poland has to be replaced. In Rome the marble of Trojan's column was defaced and many such stone monuments were moved indoors. The distinctive cast iron spire of Stockholm's

13th century Reddhanholm Church, was corroded so much that it had to be replaced in the sixties. The restoration of state's bronze monuments in USA cost several million dollars. Every major tower of London became corroded.

Sulphuric acid reacts with limestone or marble buildings to form calcium sulphate or carbon dioxide and the surface peels or dissolves in the rain.

India's pride the Taj Mahal, the famous historical monument in Agra is being polluted by the emissions of Mathura oil refinery which is 40 kms away from it. However, there is controversy over the pollution of the Taj by the emissions of this refinery. According to Mr John Mc Cormick, the author of the book *Acid Earth*, 'the pollution caused by refinery would destroy the Taj.' An Italian consultant firm brought in to advise on the environmental consequences of the refinery, concluded that sulphur dioxide emissions from the refinery would be insufficient to damage the structure and that the existing decolourisation of marble and sandstone in monuments in the vicinity was due to fungus and algae rather than pollutants.

Acid Rain Affected Areas: Global Distribution

As mentioned already, thousands of lakes and several large rivers in Scandinavia have been polluted causing death of fish and other aquatic fauna and flora.

Drinking water was contaminated by leaching copper from water pipes by acid rain in Sweden.

Historical monuments and structures were corroded by pollutants of acid rain in US, UK, Italy, India and several other countries.

Large-scale damage to forests occured in several countries. In Germany, Switzerland and Sweden about half the natural forests are dead or dying and 60 per cent of all trees in Britain show signs of decay.

Acid rains adverse impact has also occured in China, Japan, Australia, Mexico, Brazil, Venezuela, Nigeria and India.

Eighty per cent of lakes in Norway and a quarter of these in Sweden are acidified loosing aquatic life and 20 per cent of US lakes are sterile.

Recent measurements in Scandinavia, Central Europe, Japan have shown PH values of below 5 or even 4 over large areas. Values as low as 2.1 are shown in North Eastern US and 2.4 in Scotland.

Human beings are subjected to diseases due to impact of acidified smog and acid rain pollutants in mny countries like UK, Japan and Scandinavia.

Today industrial regions suffer much due to acid fall-out. Acid rain covers the whole of Europe, Canada, USA and Scandinavian countries. By 1982 there were 18,000 acid lakes in Sweden. The lakes in this region were known for their diversity in flora and fauna. In Canada over 40,000 lakes have been permanently polluted by acid rain.

Acid Rain Impact an International Problem

The pollutant gas emitted in one country after reaching the atmosphere travels thousand of kilometres depending upon direction, time and speed enters neighbouring countries and causes damage there. As such it has become a cause of friction between countries and it has become an international problem.

It was a Swedish soil scientist in 1960 who first traced the cause for elimination of fish and other aquatic organisms from Sweden, due to acidification of water bodies on account of emissions from UK and eastern Europe reaching Sweden. The Swedish scientist Swenti Oden's sustained campaign on acid rain and its causes and possible effects, kindled interest. It was raised by Sweden at UN conference in 1972. Oden's report warned of a chemical desert engulfing entire Europe and other parts of the world. The acid rain problem became an international issue. On the strength of Oden's arguments the Organisation of Economic Cooperation and Development (OECD) in Paris set up a Long Range Transport of Air Pollutants Programme (LRTAP) followed by a European monitoring and evaluation programme under the auspices of Economic Commission for Europe—all to monitor sulphur deposition rates.

Slowly tension developed between UK and European countries and between US and Canada also. By 1983, the sulphur deposits amounted to 84,000 million tons of which 20 per cent was from other countries. But 92 per cent of Norway's, 82 per cent of

Swden's 77 per cent of Netherland's, 64 per cent of Denmark's, 58 per cent of Belgium's and Poland's sulphur deposits originated from UK. By 1976, Britain had to confer its complicity in fowling the atmosphere over Norway and Scandinavia.

The per capita energy consumption of US is higher and its sulphur emissions may be about 22 x 10^6 tons by the year 2000. These pollutants affected about 2/3rd of land area in the US itself. But it was Canada which received most of its acidic gases which turned its lakes barren and forests dead. Acid rains thus became a bone of contention between many countries because winds carry pollutants across international boundaries. As mentioned already the pollutants emitted from sources in the industrial heartland of mid-western US and Central Canada regularly fall on the comparatively less polluted areas of SE US and Canada. It has been a long standing disagreement between Canada and USA and the Scandinavian, western and central Europea countries. While Canada sends half a million dangerous emissions to US, it receives 2 million tons in return. Norway and Sweden claim that they receive emissions from US. Similar charges have been made by France and Germany against each other.

Remedies/Measures Suggested to Control Acid Rain

One of the ways to control the effects of acidity in the short term is to clean up air with dumping vast quantities of limestone into soils, acid waters and adding fertilizers and nutrients to forest soils.

Prof. Rodhe, has suggested a multiprolonged approach in the Indian context since the potential for acidification is increasing manifold. He suggested that coal based industries should be decreased and coal with less sulphur content is to be used. To enforce anti–pollution measures latest technologies in all industries are to be adopted. The industries are to be distributed in a staggered manner so as to avoid pollutant concentration in any localities. Catalytic converters in automobiles to reduce nitrogen oxides and hydrocarbons have to be fixed in automobiles.

Acidification remedial measures were also discussed in Stockholm conference 1982. In 1993 the executive members of European community including Finland, Norway, Sweden proposed 30 per cent cut in sulphur emissions before 1993.

The concentration of sulphur varies from 1 to 5 per cent in coal, which through its oxides primarily contribute to acid rain phenomenon. Removal of sulphur from the coal is the best solution but the chemical engineering is not successful. There is a clue that certain microbes may achieve what the engineers have failed to do as sulphur in coal occurs in organic and inorganic forms.

Technology now exists however for limiting the emissions from power stations, factories, and motor vehicles. But it is expensive and uneconomical. Many power stations have turned to low sulphur coals and are installing emission control devices. The filter system commonly used is the sulphur scrubber or fuel gas desulphurisation vent. This uses calcium hydroxide which is obtained from limestone (unfortunately CO_2 the green house gas is liberated in this process) scrubbers are expensive. The British government has imposed tough emission control on Scottish power generating companies.

To protect forests and lakes, emergency measures have been taken in Europe such as spreading lime on them, fertilizing damaged trees with calcium, potassium, magnesium and so on.

Motor vehicle emissions are to be cleared by catalytic converter which is fitted to the vehicle exhaust. Catalytic converters which absorb CO, nitrogen oxide and CO_2 are compulsory in US and Japan. All preventive measures are expensive. The environmental gain has to be balanced against costs and detrimental factors that may arise such as excessive quarrying of limestone for the scrubber process.

Every year global atmosphere receives 130 mil tons of SO_2, 20 mil tons of CO_2, 97 mil tons of carbons and 53 mil tons of nitrogen besides other chemical pollutants.

Experts believe that if we use the latest technology available such as fuel-gas desulphurisation technology as is being done in Europe with existing power plants (coal and oil) an investment of 72,910 crores per annum is required to counter the acid rain by 2020. But if we use basic control technology (emissions from large point sources, controlled by domestic technology such as limestone injection) then our country may require 26,545 crores every year for pollution control measures, says T.S. Panwar of the Centre for Environmental Studies, TERI.

In India, 70 per cent of coal is used by power stations. If India develops a technology to provide a desulphurised coal to power stations, a major part of the acid rain problem would be solved.

National Environment Engineering Research Institute (NEERI) of Nagpur developed a technology to remove sulphur from coal. It has been demonstrated in laboratory. There is need to conduct a pilot plant study before making the technology available for mass use. But there are reports that Western Coalfields Limited and the Union Ministry of Environment and Forests has turned down the NEERI's proposal to find the setting up of a plant project.

According to a Supreme Court order all the refineries will have to produce diesel with sulphur content less that 0.25 per cent by the year 1999. The number of diesel vehicles are growing due to large difference between diesel and petrol prices. A large number of gensets are diesel oriented. The increase in use of gensets are due to shortage of power and power cuts.

Several oil companies in India are setting up hydrodesulphurisation plants to produce low sulphur content diesel which would be less polluting to the atmosphere. So long as the government keeps the diesel prices low, all these efforts become in vain.

To tackle the problem the government needs to determine first the acid rain load for different areas of agro-climatic diversity of the country. An acid rain model that can predict the amount of pollutant-loaded rain water falling on a particular area in a given time frame should be worked out based on which one can determine the probable damage caused to human beings and land. Thus, damage caused by acid rain should be taken up using the available control techniques.

It is quite interesting to note that advanced countries like Japan, South Korea, and Taiwan with higher industrial production than India have a low atmospheric pollution load mainly because they adopt appropriate pollution control measures.

Current policy operations for controlling the problem are residual improvements in energy efficiency, switch over to low sulphur fuels like natural gas, greater use of renewables, major cut down and removal of sulphur from crude oil distilleries such as diesel and fuel oil and finally the widespread use of strict pollution control devices in all sectors of the economy.

In Europe and North America the threat of acid rain was severely dealt by heavy spending on SO_2 abatement and rapidly cutting down the dependence on coal by shifting to natural gas and nuclear energy.

Scientists are of the opinion that the acid rain can be controlled if India adopts pollution control technologies. However, strong political will is needed to recognise the urgency of the problem.

In the absence of even low investment in the basic control technologies or domestic technology as it is known in India, China and Pakistan, the situation could deteriorate to a point where the concentration of acids in acid rain could increase tremendously causing burns and destroying crops on contact.

Chembur in Maharashtra experienced acid rain during the 1970s and 1980s because the industries in this region had been using fuel with a sulpur content.

Controlling industrial pollution pays in the form of healthier population, high agricultural production and abundant plant resources.

Thus, acid rain is a consequence of industrial development without regard to environment. The Swiss, Germans, and Austrians have rapidly become conscious of their environment. The reductions made so far may not be enough to reverse the already advanced destruction of forests, rivers and lakes. But south-east Asia and Latin America may become problem areas since they are developing rapidly and their energy demands are growing by leaps and bounds. But there is no sign of any constructive plans to avoid the mistakes made by the west. China, India, Malaysia and Brazil may be rushing head long into the kinds of problems that have become only familiar to west Europeans and North Americans. These countries should bear in mind the problems faced by developed countries while adopting acid rain controlling measures.

Acid rain pollution is a global problem which can be tackled with international understanding and cooperation between several countries.

6

NOISE: THE NERVE-ROCKING NIGHTMARE

Noise has became a part of urban living and modern civilization. It has not been recognised as a suffering and the modern man has adapted to it. The recognition of noise as a pollutant is rather new because its negative effects are subtle. It is easy to recognise a heap of garbage or dirty stream but not the adverse impact of noise. In fact it affects a vital organ of human beings. During the earlier period of industrial revolution, noise was confined to factories, but with increase in urban transport noise spread to every corner of the cities and the noise pollution is growing faster than the growth of population.

Types of Noises

All technological innovations like jet planes, railway engines, vehicles, loudspeakers, industries, generators, grinding mills, and several others are the sources of noise pollution. Many international organisations, such as international civil aviation organisation, organisation for economic cooperation and World Health Organisation have carried out several studies to identify culprits of noise pollution. Various attempts have been made to identify sources of noise to adopt methods of abetting and controlling them.

Measurement of Noises

Noise is measured in terms of relative units of energy or power on logarithmic decibel (dB) scale according to the response of the ear. Zero dB is the threshold of hearing and 130 dB is that causes pain for human beings. Normal conversation is about 60 dB and the ticking of a watch is 30 dB. The noise made by aeroplane is about 100 dB.

The annoyance level for most practical purpose is determined by measuring the sound pressure level on the A-weighted network on the sound level metre. This is a filter circuit which gives output corresponding to equal loudness curve of 40 phons. This has been accepted for measurement at all sound levels. These measurements are indicated as dB(A).

Several surveys in different areas have been made to find out the intensity of each type of noise, which is expressed in terms of dB. Dr. Henring Evan Geerke has established correlation between population and density of average noise level. According to a survey by Galloway if the population is 10,000 persons for one square mile, the average day and night sound level is 52 dB and it reaches an average of 72 dB with a population of 1 million. Large metropolitan cities become centres of noise pollution.

Besides factory noises, in large number of cities people are exposed to jet plane noises. A jet plane at take off causes a noise of 140 dB. Continuous exposure to a noise of 85 dB causes damage to hearing.

In India the situation is more pathetic. Some years ago the National Physical Laboratory, New Delhi conducted studies on traffic noise in major cities of India. The studies revealed that street noise during the day in cities like Delhi and Mumbai was intolerable as high as 90 dB and it was never below 60 dB. Similar conditions are found in many cities of India. In busy localities during the day the level exceeded 100 dB. In Mumbai the average day and night noise level is 75 dB, the highest being 105 dB near the airport. The main city area continues to remain noisy even during the night, whereas the noise level in the suburbs subsides after 11 a.m. Besides the traffic problem, the residents of Kurla, Ghatpur and Santacruz suffer from even more critical noise problem from approximately 50 jet planes daily. The city is not free even on Sundays and holidays. During holidays many areas of the city are disturbed by record players and loudspeakers used in social functions.

Recordings taken during Ganesh festival procession showed maximum level of noise of 106 dB. Any one standing about 5 metres from the bursting crackers will be exposed to 128 dB. A level of 120 dB is designated as the threshold high discomfort capable of causing serious damage to ears.

While the WHO has prescribed a limit of 55 dB during the day and 45 dB during night, noise levels exceed the limit in many cities. The quietest period in many cities is from 2 a.m. to 5 a.m. Between 10 a.m. to midnight noise levels are relatively high.

The permissible limits of exposure in industry specified by the Occupational Safety and Health Act (OSHA, 1970) is 90 dB for eight hours continuous work. However, Sweden, Denmark and Russia set their own standards of 85 dB. The aim is not to stop noise pollution which is not possible but to offer to society an environment which will be found practical as well as optimal for daily activity of society. Nearly 30 to 35 per cent of employees exposed to noise levels of greater than 90 dB in various industries suffer hearing loss.

Noise levels of 90 dB have been measured at a distance of 10 metres from the railway track. This level has been recorded in busy railway stations. During festivals like Diwali noise levels go to 80-100 dB and loudspeaker noise levels reach 80-95 dB.

Impact of Noise Pollution on Human Beings

As mentioned already noise is a slow agent of death. An American environmental expert predicts that if the noise level continues at the present rate most people living in metropolitan areas will become deaf by 2050. Every year noise pollution is increasing in quantity and variety and most people adopt to it. Noise affects nervous system and causes damage to ears. Besides hearing loss, people might suffer from various other health hazards such as sleeplessness, communication disturbances, etc. Unwanted noises like barking of dogs, hawkers, beggars often continue some times even until late hours adding to the pressures of city life.

During the last 50 years, a number of experiments were conducted on animals and human beings to study psychological, biological and physiological effects of noise. As such, the ill effects of noise can be divided into three categories, i.e., physiological, biological and psychological. In all three categories effects are varied, direct, indirect, short-term, long-term, serious and minor. All these effects depend upon the frequency and duration of exposure, type and other environmental factors.

Depending upon the frequency and intensity of exposure an individual's hearing can either be impaired or lost completely and permanently leading to deafness. Loud and continuous noise can cause rapid beating of the hearth, constrict blood vessels and lead to pressure of blood circulation and even cardiac disturbances. For example, bad noise causes peptic ulcers, gastro intestinal problems, allergy and nervousness.

As far as biological effects are concerned it has been found that in female rats noise can cause infertility, malfunctioning disorder, and even confused behaviour. Long-term biological and genetic effects of noise are yet to be established.

Regarding psychological effects, noise creates communication disruption, frustration, sleeplessness and people easily get tired and irritated. It has been established that the number of accidents in noisy factories are higher than in quiet ones. Many social conflicts and lack of cooperation have been attributed to the unwanted noise which keeps individuals irritated, upset, unbalanced and fearful. Very few people escape harmful effects of noise. Even serious adverse consequences such as paranoid delusions, suicidal and homocidal tendencies, hallucinations and even psychotic behaviour may occur.

The strange characteristic feature is that one does not know when he starts loosing the capacity to hear. It is only when he reaches a significant stages that one realises that he has lost hearing and at that time it is too late even for the best medical practitioners to retrive what is lost.

Hearing levels in urban and rural individuals reveal urban people are poor in hearing when compared to rural people. While the normal level of hearing at low and high frequencies should be less than 20 dB, urban people were below normal (25 dB) and the industrial employees were still poorer (40 dB). This is due to frequent exposure to noise in the case of urban and industrial workers. One need not be surprised if future of urban people in cities exhibits a serious hearing problem or even if babies are born deaf.

According to a survey on occupational hazards, short exposures to intense noise can shift upward the hearing threshold while prolonged exposure or intermittant exposure for longer period produces a damaging effect on hearing resulting in a permanent threshold shift. According to another study 1.3 per cent of industrial workers complain of tinnitus for noise exposure levels upto 80 dB. 11 per cent complain of tinnitus at 80-90 dB level. The number keeps on increasing till 110 dB level is reached when complaints become universal. Audiometric tests conducted for weavers shows that exposure to noise levels of 102-104 dB for 4 to 8 hours can result in their temporary hearing loss.

Dr. Samuel Rosin, a pioneering researcher, while addressing a meeting of the Accoustical Society of America in 1969 warned that we have millions with heart disease, high blood pressure who need protection for additional stress of noise. If a disorder such as arterioclorosis or coronary heart disease is present, noise exposure can endanger health, protection from any noise is important. As mentioned earlier exposure to 75 dB of noise for more than 8 hours can cause deafness, hypertension, disturbance in sleep. Speech interference and stress are other effects.

According to WHO one can have sound sleep at an average of 35 dB sound pressure.

Measures to Control Noise

Noise is an important environmental problem. 70 per cent can be controlled by designing machines, human behaviour and structure of the city. Noise pollution is not on the agenda of planners or legislators and people give little importance to it. Of all the pollutions, noise pollution is most neglected, consciously ignored and accepted. Technically from accoustic point of view there are three methods for controlling noise. These are: regulating noise at source, interrupting noise in transition and protecting the receiver. Many industrialised countries of the world have passed legislations to control noise. In fact, the first Noise Control Act in US was passed in 1955 but the US. Noise Control Act of 1972 is a major breakthrough to eliminate excess noise at every level. The environmental protection agency which passed the Act has information about permissible levels of noise and set noise emission standards for products which are major sources of pollution. It advises on setting standards for noise emitted from aircraft. Its main concern is to protect health and safety of citizens. Most importantly, the manufacturers and importers who do not confirm are subjected to fines upto 25,000 dollars per day for each violation and imprisonment upto one year. Several types of activists and programmes to control noise pollution have been taken in large cities of US. India also should enact similar legislation recognising the danger of noise to health and welfare to the people in urban areas and provide a noise free environment.

The following steps could be taken to control noise pollution:

1. Identify major sources of noise pollution such as transportation vehicles, construction, ceremonies, etc.
2. Specify locations and number of people affected by noise.
3. Examine health hazards caused by noise pollution.
4. Set emission standards for all sources of noise which are feasible and necessary.
5. Impose strict penalties on offenders.
6. Set up government machinery to check noise pollution.
7. Prepare extensive network of information system for the public for awareness and education.
8. Provide funds for research and development in noise control technology particularly transportation machinery, electric and electronic gadgets.
9. There should be political motivation.
10. See that local pressures are systematically mounted against noise menace.

Controlling noise at source is the most economical and efficient method. It is essential to new technology to confirm to the standards of safety. Those which do not confirm to the standards of safety should be banned. Controlling noise pollution is an expensive proposition. A great deal of money is required at every stage from individual to the government. The public should be willing to opt for less noisy products. The polluter should pay to control noise pollution. In any case a great deal of finance will be required not only for designing machines but also for research and development for pollution free technologies.

Many accoustic problems result from the haphazard unscientific and mistaken designs of our cities and buildings. Some noise is directly proportional to congestion of buildings and population. Planning must emphasise sparse and dispersed settlements. The location of buildings should be properly planned. A hospital or school should not be near theatres or shopping centres. Placing a residential area at a hill, near a highway, or a rail road is a wrong choice. On the other hand the hill can be used as buffer between the two. Principles of accoustics should be used for relative quiteness while utilizing the space. The city of Gothberg reduced noise by simply redesigning a thoroughfare in relation to control business area. Concurrently the transportation system was so designed that the vehicle volume was reduced by half in a congested area resulting in the reduction in noise.

No amount of legislation or technology can eliminate noise unless the citizens are sufficiently educated for wanting a noise free environment. As such, citizens participation is essential to have a noise free environment.

In the United States, people are so conscious about noise pollution that many communities have developed their own noise advisory councils and community noise control ordinances have been passed recently to plan for a desirable environment. To quote from the model community noise control ordinance of September 1975 each community has its own set of environmental, health, economic and other goals it wishes to attain. Each community has its own configuration of noise sources and their impact which it wishes to control. The level and extent of speech control is fully within the purview of local decisions. Local decision requires citizen involvement which means requisite knowledge. Education concerning the environment and its problems should become widespread all the way from primary schools to universities. There should be absolute awareness among the people that noise is a potential hazard which causes several ailments, besides severe deafness.

Besides all these to cut noise levels, public awareness about their ill effects should be created through radio, TV, press, posters, lectures, symposia and demonstrations. Also administrative measures to ban the use of loudspeakers at night, limiting the amplifier power can be adopted to keep the noise level low.

Noise made by vehicles can be reduced by banning horning and streamlining traffic flow. The rules formed for checking noise pollution should be strictly implemented.

Industrial noise can be controlled by using materials that can absorb sound. Workers should be provided with ear defenders or plugs. This is the cheapest and best method. Their working hours should be adjusted to ensure that they are not exposed to noise beyond the acceptable limit. Audiometric tests must be done annually.

Rotation of workers in various sections where possible to give an exposure not exceeding the permissible level is an alternative. Another way of solving the problem is to detect the persons likely to loose hearing capacity for a given exposure level and duration and withdrawing those persons from noisy places. If the noise level at the work place and duration of exposure are known, the likely hearing decrease level can be found easily from application nomograms. Once the danger is established, suitable steps can be taken to place individuals in appropriate work sites. Individuals susceptible to noise pollution can be protected by taking proper measures. Hearing ability decreases with old age. It is accelerated with exposure to noises.

To check and punish law breakers certain laws have been incorporated in Indian Penal Code and Motor Vehicle Act of 1951. Despite all these, the problem of pollution continues. Lack of seriousness in application is one of the causes, other contributing factors being non-availability of measuring equipment.

Recommended Noise Levels by WHO

On the basis of extensive research into human response and preferences, the World Health Organisation (WHO) has recommended noise assessment criteria in the form of noise exposure limits which are as follows:

TABLE 6.1

NOISE EXPOSURE LIMITS RECOMMENDED BY WHO

Environment	*Recommended Maximum Level dB (A) Leq.*
1. Industrial/Occupational community (urban)	75
(i) Day time	55
(ii) Night time	.45
2. Indoor	
(i) Day time	45
(ii) Night time	35

The criteria has been in equivalent sound dB(A) Leq.This measure is defined as the average A weighted mean energy of noise level over the measurement.

Standards to control urban pollution have been also laid down. The recommended ambient noise level limits for various areas are as follows.

TABLE 6.2

RECOMMENDED AMBIENT NOISE LEVEL LIMITS

Area	*Day time 6 a.m. to 9 p.m. Leq dB (A)*	*Night time 9 p.m. to 6 a.m. Leq dB (A)*
Industrial area	75	65
Commercial area	65	55
Residential area	55	45
Silent zones	50	45

(Silent zones are areas upto 100 metres around hospitals, educational institutions and courts)

Standards for maximum allowable noise dose (occupational noise loss) for individual workers have been suggested under environment (protection) Act-90 dB (A) for eight hours shift with the time limit to be halved for every 3 dB (A) increase exposure to continuous or intermittent should not exceed 115 dB (A) and exposure to impulse noise should not exceed 140 dB (A) peak accoustic pressure.

We need to develop our own standards for acceptable noise limits taking into consideration economic growth of the country. Strict imposition would impede over the economic growth of the country. As such, socio-economic and technical factors should be considered while attempting to stipulate any standard. Adoption of standards of permissible limits recommended by World Health Organisation in a country like India with its huge population and congested traffic may be unrealistic or even ineffective. Scientists and law makers must putforth their efforts to establish for setting an acceptable limit which is practicable in our country's socio-economic setting. Therefore, noise control needs careful planning taking into account the environmental and socio-economic conditions of an area.

7

SLUM MENACE: A THREAT TO THE ENVIRONMENT

Be it Ever so Humble, This is Their Home

Squatters' settlements form a major threat to environment in third world countries besides untreated and non-recycled industrial wastes and effluents, domestic waste and garbage dumps, non-pollution free vehicles plying on the roads, and lack of sanitation and drainage facilities. In the third world countries, urban population is increasing at a faster rate. While urban population growth rate is expected to raise from 450 mil to 1.0 billion in developed countries, it is expected to rise from 300 million to 2.0 billion in third world or developing countries between 1950-2000 AD. Out of 25 major cities in the world 19 cities are in developing countries. In India, the urban population which was 200 millions in 1981 is expected to reach 340 million by the end of this century. Due to rapid growth of industries and other activities there is large scale migration of people, particularly from rural areas in search of occupation, due to which the cities and towns are becoming congested and over crowded.

Among the migrants to urban centres, squatters form the major component. A squatter is one who takes over land or a house or building and occupies without the legal authority to do so. The squatters' settlements are mostly huts or low cost houses built with tins or tarpaper or with leaves of palmyra, etc. built mostly in most of the government or sometimes private vacant lands or even on pavements of roads unauthorisedly, by occupying them. Most of the squatters are poor and are well below poverty line and living in very unhygienic conditions.

Mexico for example has large number of such settlements in a hill side dump. In 1981, 40 per cent of the residents of the third world lived in squatter colonies. Before the end of 20th century, 3 out of 5 urban people are expected to be squatters. In India huge squatters' settlement colonies have developed in big cities like Mumbai and Calcutta.

These cities have now become the biggest breeding ground for slum dwellers. Even in other cities huge colonies of these settlements are a common feature. The vast majority of the squatters occupy these lands out of sheer necessity, since they are unable to obtain land or house for a reasonable price. Living outside the city is not feasible to inaccessibility and high cost of transport to reach their workspots. The squatters' settlements in general have no facilities of sanitation, drainage, garbage disposal. The congested and overcrowded conditions lead to the spread of diseases and contribute to high rates of death and chronic illness. Besides the traditional materials used by these slum dwellers for working and heating cause much pollution in addition to the unhygienic conditions prevalent in those localities. Some of them may be exposed even to indoor pollution at night. These slum areas have also become breeding centres for all kinds of criminals in several cities.

In Mumbai 50 per cent of all housing is made of shanties. Although 15,000 dwellings are built in this city every year by both public and private investors, the government estimates, that more than four times that number are needed simply to meet current needs. In Karachi of Pakistan 2.6 million persons i.e., 40 per cent of its total population live in squatters' settlements in 1986. Likewise in Bogota of Columbia (S America) illegal housing accounted 50-60 per cent of residential houses during the last two decades.

Lack of affordable financing and material for building homes, low incomes, high level costs, inability to secure loan at low interest rates, not only prevents poor people from buying or building a home but also hinders their attempts to make slum areas safer and healthier. They are often reluctant to invest in household or community improvement.

To maintain a healthy environment in urban areas, the first and the foremost thing to do is to shift squatter slum dwellers from all vacant lands, including pavements of roads by providing alternate sites for rebuilding their dwellings. Necessary loans for low cost housing and subsidised transport facilities should be provided to reach their workspots.

To maintain urban ecosystem the optimum size of a city with reference to population shall be decided on the basis of urban environmental parameters like availability of vacant land to support vegetation for absorbing the carbon dioxide (CO_2) contributed by the people, and the transport vehicles, so as to maintain normal levels of CO_2 in the atmosphere. Any increase in population, will increase the transport vehicles and the combined effect of both would increase the CO_2 levels n the atmosphere as such. More vacant land is needed or the population of a city or town should be reduced by encouraging migration which is possible only by transferring some of the industries and other secondary and tertiary activities. Secondly the optimum size of city should also be based on the drinking water availability, particularly ground water.

If the vacant lands are occupied by squatters' settlements, they are devoid of natural vegetation resulting in the rise in CO_2 levels in the atmosphere. The increase in CO_2 level in the atmosphere raises the temperature which warms up the sea and the rise in sea level would submerge vast coastal areas. Many of the coastal towns and cities will be affected largely with the rise in sea level in the coming fifty years.

Secondly, the unhealthy and filthy conditions created by the slum dwellers in the heart of the cities and towns, causing environmental pollution could be also prevented. As such, all the vacant lands within the city should be free from squatters' settlements and should be converted into green belts. Besides, transferring some of the industries, a large number of other activities can be shifted to small towns and big villages. This will not only check further migration of people from far off rural areas to large cities but also result in migration of a large number of slum dwellers from these big cities in which all vacant lands can be converted into green belts. Transfer of these activities to several growth centres selected on the basis of environmental parameters may develop into small cities and towns. Rise of more cities and towns and their even distribution would also generate more urban-rural interaction.

Moreover, it solves many problems like housing, power and water supply which are dominant in the present unevenly distributed large cities which have almost become slum cities due to large-scale immigration of population.

As such decentralisation of industries and other activities by selecting growth centres based on environmental parameters like availability of vacant land for vegetation, barren land for housing is the only solution for environmental preservation in urban areas. These growth centres would eventually develop in to small cities and towns providing employment to a large number of migrants.

If environment is not protected the whole mankind would perish. Hence people should be awakened about this and top priority should be given to preserve environment at any cost.

The seriousness of the problem should be understood properly by the decision makers but should not be politicised.

8

COMBATING DESERTIFICATION

Desertification which is a process of land degradation, loosing all vegetation cover from its surface particularly due to adverse impact of man, has now become the worst environmental hazard on a global scale.

Cold Deserts

Any area is called a desert if its surface is permanently devoid of any vegetation. Deserts are formed due to abiotic/ physical or natural factors like low rainfall, snow coverage or due to biotic influence as caused by man and animals. Deserts caused by natural factors are two types, the hot and trade wind deserts like Sahara, Western Australia, Atacama, Kalahari which are found in the Western margins of the continents between 20° and 30° latitude. These deserts were formed due to high pressure belts and offshore trade winds blowing from land to ocean throughout the year. Since these winds blow from land to ocean always, they do not carry any moisture to give rainfall and remain as dry winds causing deserts. Hence the annual rainfall in these deserts is less than 160 mm and in some places rainfall occurs once in five to six years. As such these regions cannot support any vegetation which needs sufficient rainfall. In India, the Thar desert of Rajasthan is a good example. Cold deserts are areas covered with ice throughout the year preventing the vegetation to grow. These are located in polar regions of the world.

Besides these two categories of natural deserts, there are deserts being formed by animal and human interference. This kind of land degradation leading to desert formation is taking place on a large scale in dry lands of arid, semi-arid, and subhumid types by human interference mostly. This kind of desertification is the destruction of the biological potential and can lead ultimately to desert-like conditions. There is a widespread deterioration of ecosystems with the adverse impact of man.

Over the land surface of the earth 6.1 per cent billion hectares, nearly 40 per cent of land is dry. Of this 0.9 per cent billion hectares of area is under natural hyper arid deserts. The remaining 5.2 billion hectares of area is dry land which is classified into arid, semi-arid and sub-humid types. Out of this 5.2 billion hectares of dry land, 3.6 billion hectares, i.e., 70 per cent is affected by desertification and it also affects 1/6 of world's population. More than 100 countries are affected by this desertification. Out of the total area affected by desertification 63.5 per cent is in developing countries and the rest is in industrialised countries.

Land degradation leading to desertifiction is caused due to deforestation/overexploitation of forests, shifting cultivation practices, overgrazing of animals, overexploitation of ground water, heavy chemical inputs for intensive crop culture, use of pesticides causing toxicity in soils, wrong cultivation practices, abandoned mining areas, industrial effluents, industrial hazard wastes etc.

About 73 per cent of range lands with low potential for human and animals carrying capacity have been degraded. These lands may have considerable capacity to recuperate and regain their potential productivity if properly managed. Large degraded range lands are in Asia followed by Africa. Much of deforestation is caused in Himalayan region in India which is loosing forests at the rate of 2.8 per cent annually.

Rainfed crop lands which constitute 47 per cent of total dry-lands are affected by various processes of soil degradation — mainly water and wind erosion, depletion of nutrients and urbanisation. Consequently 3.5 to 4.0 million hectares of rainfed crops are lost every year throughout the world's dry lands.

This is indirectly affecting the best range lands in reducing their area by cultivating them to compensate.

There is also degradation of 43 million hectares of irrigated crop lands amounting to nearly 30 per cent of their total area in dry lands, which usually have highest agricultural potential and greatest densities when well managed. Largest area of degraded irrigated lands formed through processes of water logging, salanisation and alkalanisation are found in Asia.

In terms of degree of desertification, severely affected area accounts for 46.4 per cent which is highest when compared to very severely (8.8 per cent) and moderately (44.8 per cent) affected areas in the whole world. Very severely affected area (11.5 per cent) is highest in semi-arid areas and severely (80.7 per cent), moderately (80.0 per cent) affected areas are highest in arid and subhumid areas respectively.

Rehabilitation Measures

Agricultural expansion to marginal lands frequently resulted in rapid land degradation, with a subsequent decline in production. Hunger for land often causes agricultural encroachment by marginal farmers into marginal drylands. This demonstrates unwise land use leading to poverty. Unless adequate livelihood is created elsewhere it is not possible to stop agricultural encroachment on marginal lands and consequently desertification.

The United Nations plan of action to combat desertification has suggested several preventive and corrective rehabilitation measures like improved land use systems, developing and introducing appropriate agricultural and pastoral technologies that are socially and environmentally acceptable and economically feasible, developing appropriate water resources, introducing improved water management systems, reclaiming severely desertified lands for productive use or for protecting effectively for natural rehabilitation.

Protecting lands that are not degraded or slightly degraded and maintaining their productivity is considered a first priority in combating desertification and this is economically viable and partially feasible. The second priority is, applying corrective measures and sustaining the production of drylands that are moderately degraded at present. The rehabilitation of very severely degraded dry lands and their return to productivity use is considered to be third priority action.

For implementing all these measures to combat desertification the total estimated cost is about 22.4 billion dollars per year.

9

URBAN PLANNING AND ECOLOGICAL BALANCE

Where is the Good Earth?

The Urban ecosystem is characterised by the interactions and interrelationships between the urban environment created by man, the most powerful member of biotic community, the biotic environment created by other groups of biotic community i.e., plants and animals, and abiotic/physical environment comprising relief, soils, land forms like deserts, water bodies like rivers, lakes, etc. and climate (temperature, rainfall, etc.,) in an urban area which is a spatial areal unit of earth's surface and this is referred to as a town or a city depending upon the size of its population.

The urban environment in an urban areal unit, i.e., a town or city is characterised by a number of man induced secondary or tertiary activities like industry, construction, transport, trade, commerce, etc. Most of the working population is engaged in these activities. In an ideal urban ecosystem the stability or equilibrium between and within urban, biotic and abiotic environments is maintained. But this equilibrium which in other words called ecological balance rarely exists in urban settlements and this unfortunate situation is due to over interference and exploitation of the biotic and abiotic environments by the man for expanding and intensifying his urban environment, and this eventually leads to environmental degradation.

Even though greater ecological imbalance exists in urban areas, the urbanisation is considered to be a development process. The movement of people for various occupations is an intimate part of development representing spatial dynamism of development process. Urbanisation is changing the socio-economic and political structure of the countries as well as affecting the environment. It is a dominant demographic process in the 20th century. The number of people living in towns and cities increased from 600 million in 1950 to 2.1 billion in 1989. By the end of 20th century, half the humanity is expected to be living in urban areas. By 2006, about 3.6 billion people may be living in cities and towns covering only 1 per cent of earth's surface.

India one of the best examples in this respect. In this country in 1981, the urban population was 24 per cent of the total population i.e., 160 millions. The national commission on urbanisation has estimated that urban population would reach 340 millions by the end of this century. Housing is an acute problem in almost all cities in developing countries where low cost housing is largely not available. This is causing the rise of innumerable squatters' settlements built unauthorisedly on public land by squatter population who migrate from rural areas to urban areas is search of employment.

Since the cities are characterised by the growth of a large number of industries in developed countries, environmental degradation, polluting land, water, air, is caused by industrial wastes including hazard wastes. Besides industrial wastes large-scale consumption of fossil fuels particularly coal in industries and petroleum for transport are increasing the amount of carbon dioxide (CO_2) in the atmosphere interrupting the global geothermal carbon cycle. The rise of CO_2 in the atmosphere increases the global temperatures. It has been estimated that global temperatures could rise about 1°C-4.5°C by the end of this century. Global warming would also raise the sea level due to thermal expansion of water on heating and thereby submerge many coastal cities and towns affecting the urban ecosystem. The increase of population is also largely responsible for the rise of CO_2 in the atmosphere.

Large scale industrialisation in developed countries is also causing ozone deficiency in the atmosphere. The chlorofluoro carbon chemicals (CFCs) used in some of the industries like refrigeration, airconditioning, foam rubber, aerosal, etc., are responsible for causing ozone deficiency in the atmosphere. The ozone which absorbs ultraviolet radiation in stratosphere without reaching the earth protects all living organisms from radiation. Several nations have agreed recently to minimise the production of these CFCS and find out substitutes.

The accumulation of garbage is also polluting groundwater as well as air, both in developed and developing countries.

Environmental pollution problems due to erratic growth of industries, indiscriminate dumping of industrial wastes without recycling, flowing of industrial effluents charged with toxic chemicals without treatment are much more in developing countries even though these countries are less developed industrially. Even the biotic environment is more affected in the urban areas of developing countries. Large-scale conversion of agricultural, pasture and grazing lands, felling of trees, construction of houses and other activities are noticed in several developing countries including India. For example, the whole of north Mumbai is built on agricultural land. Even in Delhi a lot of agricultural land has come under urban use. Similarly in every Indian city large-scale conversion of these lands for construction and other activities disrupting the biotic environment is affecting the urban ecosystem. In these cities the abiotic/physical environment is also not spared. In Delhi, the flood plain of Yamuna river has been causing much waterlogging. Many rivers are spoiled by dumping the urban garbage and discharging untreated sewage. As it is, 380 million litres of untreated sewage is discharged by the city of Delhi into Yamuna river everyday and is highly polluted. From several river side cities lot of garbage and sewage reach rivers. In several coastal cities and river side cities, the coasts and river bunds are used as public toilets. Outside Indian cities, Lagos in Nigeria can be cited as best example in this respect.

The physical environment is also affected by the increasing amounts of carbon dioxide contributed by the dense population, non-polluting vehicles, large-scale consumption of oil by the vehicles to cover greater distances from residential areas to workspots, indiscriminate use of oil by affluent societies to maintain aristocratic life and indiscriminate felling of trees and destroying natural vegetation. The rise of multistoreyed buildings presenting concrete jungle landscapes in cities for accommodating large number of people are responsible for over exploitation of groundwater pitentialities. If the present rate of groundwater tapping continues, groundwater potentialities may be completely exhausted in a short time.

The increase of squatter population resulting in the growth of slums in cities and towns is another serious environmental threat. The congested and unhealthy conditions, garbage and filthy accumulations, absence of sewage and sanitation facilities, illiteracy and poverty of these slum dwellers are all contributing for the disruption of the urban eco-system to a large extent. Besides these, squatters' settlements have also become breeding grounds for all kinds of criminals.

Apart from all these, as mentioned already much damage to physical environment in urban ecosystem is caused with the increase of CO_2 concentration in the atmosphere in the cities of even developing countries due to high concentration of population and use of non-pollution free vehicles in transport.

The effect of increasing CO_2, thereby damaging the physical environment is more felt in coastal areas. The coastal deltas are affected with the rise in sea level due to thermal expansion of sea water, which is a consequence of rise in temperature. As such, coastal urban ecosystem is disrupted. It is also being affected with the interruption of its physical structures like posh hotels even well below the environmental protection zone limit of 500 metres high tide level (MTL) along the coasts as noticed in Goa for the development of tourism at the cost of environment.

To undertake any planning strategy to urban ecosystem, the urban planners should bear in mind that an urban area is an organism whose horizontal and vertical growth should be conditioned by its biotic and abiotic environments. The urban planning strategy should be such that under no circumstances the equilibrium maintained between the urban growth and these two environments should be disturbed. As mentioned already, an ideal urban ecosystem develops only when the equilibrium within and between the urban, biotic and abiotic environments is maintained. Unfortunately, urban planning in several countries is being carried on without giving much significance to environmental parameters like availability of grass land, agricultural land, groundwater-potentiality, carbon dioxide concentration in the atmosphere. Lack of this environmental approach in urban planning is the greatest setback. Therefore, the planning should be based on ecological considerations.

One of the greatest defects in urban planning is the absence of conducting a detailed urban land use survey and preparation of land use maps to know exactly the different categories of land use potential available since a urban expansion is taking place rapidly. Under no circumstances, the urban area should spread over green belts (pasture, grazing agricultural and forest land). The expansion of urban land should take place only on barren and uncultivable waste lands. If sufficient land is not available for expansion, it should be stopped.

Secondly, detailed surveys regarding groundwater potentialities in each town and city should be conducted. Keeping in view the availability of this water resource potential also, the urban expansion should be planned.

Another important parameter to be taken into consideration for urban expansion is the concentration of CO_2 in the atmosphere over an urban area. The increase in population in a city should be checked with reference to this parameter. More the people, greater will be emission of carbon dioxide by them, and higher will be concentration of CO_2 in the atmosphere. The increase in population will also increase the number of vehicles which consume more fossil fuels like diesel and petrol, etc., which in their turn contribute more CO_2 to the atmosphere due to combustion. The increase of population will also increase housing, industry, etc., which need more land and this may result in decreasing the green belts comprising pasture, grazing and forest lands, etc.

The decrease in green belt areas will lower the absorption of CO_2 by vegetation through process of photosynthesis. As such, the concentration of CO_2 will be increasing higher and higher with the decrease in green belt area and with the increase in population and occurrence of this kind of dichotomy in urban ecosystem leads to worst environmental degradation. As such to keep the CO_2 at normal levels, i.e., to maintain the ideal carbon cycle large-scale migration of people to urban areas should be checked. A balance should be maintained between the concentration of CO_2 in the atmosphere contributed by human beings and transport vehicles and the absorption of this amount by vegetation in each urban centre. Hence the urban densities should be controlled with vegetation cover for example the number of trees needed for 1,000 people and the ratio between building cover density and gardens, parking spaces, etc., have to be worked out. As such, it is essential that adequate vegetation cover should be preserved keeping in view the projected population based on the present rate of urban growth. Strict prohibition of pollution free vehicles, minimising the distances between residential areas and workspots to reduce fossil fuel consumption should be carried on. To minimise fossil fuel consumption public transport system should be expanded. The number of vehicles should be restricted to one vehicle for one family. Therefore, a city plan should be based on three environmental parameters i.e., availability of land for vegetation cover, carbon dioxide concentration in the atmosphere, and groundwater availability. Periodical monitoring of all these parameters should be done in relation to population growth and increase of number of transport vehicles with reference to environmental parameters has to be kept in mind. In addition to these, the urban ecosystem needs several other measures for its preservation.

Decentralisation of industries and other activities is another important measure to be taken. Transfer of these activities to several growth centres selected keeping in view the environmental parameters may give rise to small cities and towns. This phenomenon will check the migration of educated, unemployed persons and a large number of skilled and unskilled squatter population to major urban centres. This would minimise the traffic congestion and also minimise the CO_2 emmission both by people and number of vehicles. The indiscriminate exploitation of groundwater and other natural resources like agricultural, pastoral, grazing and forest lands can be also prevented. As mentioned already an ideal ratio should be maintained between built up area and vegetation cover. Recycling of industrial wastes, treating of industrial effluents, utilisation of garbage including night soil for generation of power in each town and city, creating percolation tanks in different parts of a city or town, providing storage facilities for rain water in each house to augment the drinking water problems, providing houses for squatter population are some important measures to be taken. There are proposals to construct plants for generating power in some cities in India utilising urban sewage and waste. Small cities and towns should be developed with all amenities including sophisticated medical facilities besides providing several job opportunities by creating several secondary and tertiary activities. This would certainly check the migration of rural population including squatters from long distances to urban centres in search of their livelihood.

Strict implementation of family planning programmes should be undertaken to check population growth which is the most vital problem of all problems. One child for one family should be propagated and this is strictly observed in China. At the present growth rate of population, India will surpass China by the beginning of next century.

Lastly there is need to the change in the outlook and lifestyles of urban people particularly belonging to affluent societies. Every urbanite should think that his or her sacred and patriotic duty is to preserve the environment for the benefit of all. One should think globally and act locally. One should use bicycle instead of car or scooter for covering short distances to conserve fuel energy resource and to minimise air pollution. One should develop a sort of feeling that he is doing this to keep his health and environment but not out of miserliness. He should not feel that bicycle is a poorman's vehicle and its usage is beneath his dignity or it is primitive and uncivilised. One should give up these wrong ideas and give up false prestige. On the other hand one should develop a consciousness that he is more civilised since environmental preservation is now considered to be the highest stage in the order of civilisation. Urban people belonging to different societies should stop overconsuming and over exploitation of food and other resources by frequently arranging big parties in star hotels even on petty occasions, which lead to not only exhaustion of resources but also production of huge amount of garbage. They should also minimise their fuel consumption by possessing one vehicle for one family. An outlook that the earth has given enough to the greedy but not to the needy should be developed. One should optimise satisfaction with minimum wants, instead of maximising output and over exploitation of resources for ever increasing wants which result not only in environmental degradation but also inflation. To minimise the garbage which is contributed mostly by the urban people, the use of gunny or nylon bags in place of polythene bags for carrying consumer goods, reusing of bottles, containers, shopping bags, reintroduction of slates in schools to reduce paper consumption etc., should also be adopted to minimise the garbage which has a major share in causing environmental pollution Reduction of paper consumption would reduce deforestation which is the potential cause for environmental degradation. Food resources should also be conserved by restricting the number of guests in social and religious functions.

To minimise the industrial pollution and to increase energy generation efficiency, developed countries should transfer technologies to developing countries. To conserve fossil fuels and to reduce pollution, non-conventional energy resources like solar energy for power generation should be utilized in all Indian cities. In cities where thermal power plants are located, coal ash waste lands should be utilized for planting trees like sisal as found near Nasik thermal power station.

Energy conservation in cities and towns particularly could be done by switching off lights and fans in offices, institutes, etc., when not in use.

Recycled industrial wastes, treated industrial effluents should be made use for other purposes. To minimise the production of these wastes modern machinery should be introduced in the industries. Ruralisation of urban society is also a vital necessity. Ruralisation means creation of rural awareness in an urban society. For example the cultivation of habits like drinking coconut water in place of soft drinks, the consumption of more fruits in place of sweets, are not only good for health but also encourage the development of horticulture as a part of agro-forestry enterprise which may divert the tribal community from the occupation of shifting cultivation to horticulture to check environmental degradation.

Urban people also should be educated about the environmental preservation, through audio-visual aids, lectures, contact programmes and also physically involving them in the implementation of all environmental programmes.

To put it in a nutshell urban ecosystem can be preserved only by transferring and decentralising a number of industrial and other activities to small cities and towns to check large-scale migration of people from rural areas, who come in search of their livelihood. To provide employment to these people a large number of small cities and towns based on environmental considerations should be planned. For this a number of growth centres based on environmental parameters may be selected.

Dispersed urban settlements system by developing small cities and towns from growth centres selected, keeping in view of environmental parameters for preventing over-concentration of people in few large cities is the only solution for preserving urban ecosystem. This will help equitable distribution of job opportunities which eventually cause decentralised migration and reduces the large influx of migrants into few large cities where they are struggling for survival.

For implementing all these suggestions to preserve urban ecosystem sufficient administrative, legislative measures are needed besides the strong political will.

Unsustainable Urbanisation has Lead to Greater Pollution

A large number of Indian cities have been environmentally polluted substantially due to unsustainable urbanisation which does not maintain ecological, biotic and socio-economic cultural environments in a city. Land, air and water of many cities have been polluted with industrial wastes, industrial effluents, garbage, toxic gases like sulphur dioxide, increased carbon dioxide in the atmosphere here and transport of vehicles which emit carbon monoxide. Besides over exploitation of groundwater and its contamination, mushroom growth of slums, large-scale conversion of pasture and grazing lands for construction, congestion due to large-scale immigrants into big city are other contributing factors.

A recent survey conducted by two experts Mr Brandon and Mr Hommann for the World Bank indicated that health costs in India due to air pollution levels are far exceeding World Health Organisation (WHO) recommended health standards.

Six of large Indian cities like Mumbai, Calcutta, Delhi.

Ahmedabad, Kanpur, and Nagpur face severe air pollution problems with the annual average levels of total suspended particles (TSP) at WHO standard. Carbon monoxide (CO) readings at traffic intersection often have been found to be excess of 5000 mg per cu. mm, and it is 50 times higher than the one hour WHO standard which is 100 mg per cu. mm. As carbon monoxide, ozone, lead and hydrocarbons are related to vehicular emissions their levels increase rapidly with a current increase in vehicular transport in the future.

As far as exposure of pollutants is concerned the final analysis of Brandom and Hommann revealed some 40,000 premature deaths in 365 cities of India with Delhi alone accounting for one fifth of them. Delhi today has severely affected from the health point of view by air pollution according to their study for the World Bank.

The study reveals that some 7,500 people die a premature death every year in the city because of air pollution levels exceeding the standards laid down by the WHO. There are an additional four million cases of hospital admission and sickness annually which require medical treatment. Besides these there are 242 million cases of minor sickness as a result of air pollution every year.

The economic costs of this ill health in Delhi ranges between 350-1400 crores depending upon the assumed cost of premature deaths and medical treatment in capital terms and this cost comes to Rs. 420-1600 per person or Rs. 2100-8000 annually per family of five members.

The data generated by Brandon and Hommann show that the total costs of health impacts arising out of air pollution are generally correlated with city size, per capita air pollution costs are significantly higher in some Indian secondary cities like Agra.

The cost of health impacts for Delhi alone, despite the fact that those arising out of sulphur dioxide, nitrus oxide, ozone, were not taken into account was 350-1400 crores a year.

According to Brandon and Hommann largest share of economic and health costs is due to water and air pollution.

The solution to this problem and to make the Indian cities habitable chiefly lies in checking the large scale migration of the people by determining the optimum carrying capacity of a city with reference to population, based on certain environmental parameters like carbon monoxide, carbon dioxide, sulphur oxide levels in the atmosphere, drinking water availability and green belt coverage in the city. There are several medium and small cities in India which are stagnant in growth and some even declassified since they are losing population. Transferring some of the industries and other activities to these through the process of decentralisation, besides starting new industries is the only solution to save this environmentally polluted large cities.

Planning dispersal urban settlement system by developing small and medium cities checks not only over concentration of people in large cities, but also saves land, air and water from pollution. Besides it helps equitable distribution of job opportunities leading to balanced growth which is otherwise also called sustainable urbanisation or the urban eco-system.

An internationally renowned agricultural scientist Dr. M.S. Swaminathan said, "The extensive environmental damage now occuring due to human greed and indifference can be halted and reversed through understanding and action."

10

ENVIRONMENT PRESERVATION AND REGENERATION

Tilting the Balance of Nature

Of late, there has been much awareness in both developed and developing coutries for environmental preservation and ecosystem development. With rapid growing population especially in developing countries like India and with technological advancement in developed countries in particular, there is much of environmental degradation and ecological imbalance. Environmental degradation is not a sudden process. It has started with the interference of primitive man in early ages. With the advancement of man's knowledge in the field of science and technology, the proces of environmental degradation has been accelerated by the modern man.

Environment in simple terms means 'surroundings'. The planet which we are living presents two kinds of environment, the physical/natural environment and the cultural/socio-economic environment. Ever since the formation of the earth till the commencement of life, there was only one environment, i.e., the physical/natural. The major components of this natural environment are:

(1) Land forms (mountains, plateaus, plains)

(2) Water bodies (rivers, lakes, oceans),

(3) Flora (natural vegetation and forests).

(4) Fauna (animals and human beings)

(5) Atmosphere (pressure, winds, precepitation, temperature, etc.).

The cultural environment on the other hand has been created by man alone to cater his basic and other needs like food, shelter, clothing, transport, etc.

The cultural environment broadly includes the following activities of man:

(i) Construction of dwellings (rural and urban)

(ii) Cultivation of crops (agriculture)

(iii) Horticulture

(iv) Irrigation

(v) Mining

(vi) Manufacturing

(vii) Power generation

(viii) Transportation (road, railways)

(ix) Recreation and worship.

Both the environments are dynamic in nature. As such rapid transformtion has been taking place from the beginning. The transformation of physical environment started ever since the planet earth separated from the solar system as a gaseous mass. This gaseous mass was first transformed into liquid and then finally into a solid state about 6,000 million years back.

Another important event of the physical environment is the origin of atmosphere. The present composition of the atmosphere, with oxygen and nitrogen mostly was not existing in the beginning. When the earth solidified from gaseous to liquid and then to solid state, some of the lighter gases like hydrogen and helium were left out and formed as atmosphere. Later on, there was evolution in atmosphere, and oxygen, nitrogen and carbon dioxide were added. The present state of atmosphere with nitrogen (78.03 per cent). Oxygen (20.99 per cent) as predominent gases had occurred due to volcanic eruptions, hot springs, weathering of rocks and plant and animal life. While the origin and evolution of physical environment is due to work of cultural environment is due to work of man. The evolution of certain physical phenomena, the origin and evolution cultural environment started with the primitive man. He must have been born in Malaysia, spread into Asia, North America and entered into Europe.

Man since his birth started interacting with nature. Even though the exploitation of nature started with primitive man, he could not do so much since he was mostly at the mercy of environment on account of his poor knowledge to exploit nature. However, his exploitation of nature started in two directions, the exploitation of nature's space and the exploitation of nature's wealth/resources. The degree of exploitation has increased with man growing in number, and with the development of knowledge particularly in respect to science and technology. The exploitation of nature by man was classified into various cultural stages as Paeaolithic age, Neolithic age, Bronze age, Iron age and Industrial age.

The exploitation of nature started with the use of fire about 1,50,000 years ago. Man discovered the use of wood as a source of energy for giving warmth when it was cold, for seeking protection of predators, and for preparing more digestible food by way of cooking. Since then, firewood became the principal source of energy and even today 50 per cent of the wood is put to use as charcoal. Firewood and charcoal gave way to coal, coal to electricity, and finally oil has become the major source of energy. Another impact of man is on nature's space for growing crops and building houses to live. There is large scale destruction of forests by tribals even today by adopting shifting cultivation, thereby causing bad land topography. This environmental degradation causes hazards like soil erosion, floods, etc. So man is destroying the natural environment by exploiting nature's space and wealth. With increase in the daily needs and comforts he has been exploiting nature's wealth particularly fossil fuels like coal and petroleum, and biomass, mostly the fuel wood.

The rural man from the beginning was exploiting firewood for cooking and heating. In India 68.5 per cent of energy is used in the form of firewood only on account of population pressure. The demand for firewood has surpassed natural regeneration and planning. There is a big gap between demand and supply. There is food to eat but there is not enough wood available to cook. It has been estimated that there would be a shortage of 137 million tonnes of firewood by 2000 A.D. Due to inadequate availability of firewood, to meet their requirement man is destroying the forest and thus causing environmental degradation. The depletion of forests is caused not only for energy purposes but for cultivation of crops by adopting shifting cultivation methods. Besides tribals, animals also contribute to this destruction by way of grazing.

Apart from forests, large scale environmental degradation is also caused in plains by destroying natural vegetation for cultivation, and construction. Even the cultivated lands are being encroached by dwelling causing a shift in land use pattern.

It was estimated that about 4.5 ml. hectares of forest area was destroyed during 1952-75 for fuel, wood and crops through shifting cultivation. The present forest area in our country is only 22.86 per cent of the total geographical area. Under the national forest policy it has to be raised to 33 per cent. Due to depletion of forests and natural vegetation, environmental degradation occurs, causing soil erosion, water logging and salinity, bad land topography, floods, droughts, desertification, change in the composition of atmosphere increasing carobon dioxide. About 90 mil. hectares of land by water erosion, 50 mil. hectares by wind erosion, 20 mil. hectares by floods and 12 mil. hectares by water logging and salinity was affected causing large scale environmental degradation. As mentioned already even animals cause environmental degradation due to over grazing. Even the natural phenomena like earthquakes, volcanic activity, tides, storms, climate changes, changes in river courses and floods also contribute for this degradation. But the environmental degradation due to these national phenomena is unavoidable.

Environmental degradation in the form of population is a widespread occurrence today due to man's industrial and technological activity. There is tremendous impact of development on environment causing surface and ground water pollution by industrial effluents, atmosphere pollution by combustion and poisonous gases, increase of carbon dioxide, and accumulation of urban sewage and waste, etc. There are innumerable examples to testify this environmental disaster like the fish affected in Kolluru lake in Andhra Pradesh due to pesticides used in surrounding agricultural fields through which drains flow into the lake, the pollution of ground water due to industrial affluents in Nacharam area of Hyderabad, Mathura refinery gases affecting the Taj Mahal and the dying forests due to acid rain in central Europe, etc.

Hence there is urgent need for environmental conservation and regeneration. In other words environmental protection can be achieved firstly by preserving it as far as possible and secondly by regenerating it whenever and wherever it is affected to maintain ecological balance. The process is called ecosystem development.

The utilisation of solar energy for generating electricity for pumping water, lighting, etc, has been undertaken by the non-conventional energy department in different parts of India.

There is urgent need to develop these non-conventional sources of energy on a war footing in view of the erratic generation of hydel power, depending upon the vagaries of monsoon, rapidly depleting coal reserves and inadequate oil resources. The development of nuclear power on a large scale is not a healthy trend due to radiation hazards. The disposal of radio active water waste material from nuclear energy plants poses a big problem and it is very expensive. Inhaling 1 mg of it causes death within hours.

While the hydro electric power plants have limitation due to location of sites, the thermal power plants involve lot of transport costs of coal. Besides, losses of electricity in transmission is another setback.

On the other hand by developing energy, from non–conventional energy sources particularly from biogas the total output of energy not only increase but also prevents water and atmospheric pollution. Energy regeneration and optimisation is the best solution to maintain the environment protection and ecological balance. This process is called the ecosystem development.

It is necessary to plan our interactions with nature and regularise our impact on nature. The objectives of ecodevelopment is the creation of environment that only sustains life but also enriches life, harmonising the works of man and nature for greater good will of all.

It does not imply that man does not touch nature for the deformation of environment since it is inevitable for the growing culture of society. The essentiality of man and nature is based on man as natural being, and nature as human society. The objective reality is the mechanism of interactions of social man and non-human nature. The naturalisation of man and humanisation of nature proceed simultaneously through interaction of man with nature.

Since interaction of man with nature is inevitable the basic objectives of development technologies for environmental preservation should be aimed to use minimum of environment space to achieve quick and rapid regeration of environment to maintain the ecosystem wherever and whenever man obstructs the ecological balance, and finally to maintain harmony between environment and man's socio-economic needs.

11

COUNTERING LANDSLIDES: A CHALLENGE OF NATURE

Man through his rapid and unplanned activities has triggered a number of landslides disregarding the dynamic stability in the ecological balance existing in nature. The increasing momentum of these activities enlarge the potential scale of damage to life and property and ecological resources.

The man-nature interaction is brought mostly along with the mountain roads. It is here that mankind emerges as ecologically dominant through a relatively short span of time.

Landslides are the worst environmental hazards which occur frequently in mountainous areas. A landslide is a fast movement of rock debris consisting of course, materials like boulders, sand, etc., along the slope of gravity. The sudden movement of rock debris along slopes causes heavy damage to human life and property, destroys vegetation cover and obstructs transport and communications.

Slope, climate particularly rainfall, snow and wind, structural and lithological characteristics of rock like, faults, joints, cracks, porosity, composition, texture are principal natural causes. But with the increase in intensity of human activities like cultivation, mining, and quarrying and construction like buildings, dams, roads, railways, and animal activity like grazing in mountain areas especially on highways after independence landslides are more frequently occurring, damaging the mounting ecosystem. This kind of man-nature interaction is brought mostly along the mountain roads.

In the present century mountains all over the world have been experiencing the increasing activities with devastating consequence. As mentioned already, the interaction is expressed through the spread of settlements, elaboration of transport activities, conversion of forests into cultivated lands and construction of vast reservoirs for irrigation and hydroelectricity in the mountainous areas.

Himalayan mountain terrain consisting mostly sedimentary rocks is the most affected region with frequent occurrence of landslides.

Frequent occurrence of landslides is noticed on the four highways of Rishikesh-Badrinath, Kalka-Rampur, Jammu-Kashmir and Mandi-Manali of Western Himalayas comprising Himachal Pradesh, Jammu and Kashmir and Utter Khand where the human activity is predominant. It is here that man through developmental activities has weakened the hill slopes. Ninety per cent of landslides occurred on these four highways between 800 and 2400 metres height because these elevations have experienced maximum human activities. This region also experiences high intensity of rainfall.

Mining and quarrying activities accelerate landsliding. Extensive quarrying causing the rocks over hanging the road increases the possibility of landslides. A landslide occurred in June 1958 near Aut along Mandi-Manali highway. Urban expansion has caused much landslide occurrence along the four highways where there has been an emergence of new market centres and towns. The expansion of urban areas increases the susceptibility of landslides because cities built on rugged slopes easily become unstable with development. The irony is that despite knowing the fragility of the areas and despite government's efforts in restricting the construction of markets in the landslide prone areas new market centres are emerging very rapidly along the four highways.

Construction, expansion, widening and repairing of roads are the most common activities which had directly generated landslides. All these activities require cutting of hill sides. The cutting process has been generally taking place without the adoption of slope stabilisation structures. Multiple landslides occurred along Kalka-Rampur highway due to hill side cutting. Here the bedrock is mainly composed of unconsolidated material. No steps have been taken so far to stabilise them. During rains the material saturates with water and slides the high due to gravity.

The use of explosives where rocks are hard causes an internal disintegration in the rock body itself resulting in landslides. The situation becomes worse when the faulted formation is disturbed by the blastings as noticed near Hanuman-Chatte along Rushikesh-Badrinath highway. The occurrence of frequent landslides on the Kalka-Simla highway is due to indiscriminate blasting done for widening the road which softens the hills.

The impact of landslides is disastrous. A single dwelling or the entire settlement may be buried. Different structures may be cracked, collapsed or may be buried under debris. They can destroy transport network, communication systems, dams, bridges and canals, and make fields useless for cultivation.

In the four highways between 1935 and 1984, 458 human beings and 2,096 cattle died. Landslides also have devastating impact on physical landscape. They can change the direction of a river, block a river creating an artificial lake, damage the soil by taking away the top fertile layers or burying it with debris and can destroy the forest wealth as well. Extensive landslides blocking the river and converting into a lake occurred near Sarbagni village on Jammu-Srinagar highway. It has blocked one of the tributaries of the Chenab river.

Landslides also affect springs. A landslide occurred 5 km from Digdaul towards Jammu had completely submerged the spring. Landslides cause enormous destruction to agricultural lands. Sometimes, the surface becomes so irregular. In extreme cases when cracks occur on the surface and land cannot be used even for pastures. In July 1961, the landslide occurred in Daduwa village in Chimoli district destroying all the terraced fields and the place was ultimately abandoned. Similarly in Himachal Pradesh the rabi crops and fruit crops worth 4 crore were damaged due to landslides.

Frequent blockade of the highways by these hazards affect the marketing of these commodities. In 1957 marketing of potatoes was affected due to landslides on Kalka-Simla highway.

Perception of this menace by the residents and consequently high degree of adjustment with these hazards is more in areas where they occur more frequently. Conversely the damage will be greater if the frequency of these hazards has been less. In other words, occassional landslide occurrence has proved more disastrous to human life and property because of the lack of adjustment of the people with the event.

The government has taken several measures to prevent the landslide incidence. Construction of walls is the most widespread structural measure adapted to control landslide occurrence at several places. These walls held back the unsupported rock material.

Since underground water is one of the main factors for landslide generation, all along the four highways drainage pipes of various sizes have been laid to drain out water from rocks into the valley thus restricting the flow on the road and to prevent its lubrication of loose rock debris from hill slopes.

Over bridging is another method used at some places along the highways to minimise losses by the landslides. The sliding material instead of falling on the highway slips down beneath the bridge thus saving the road.

Construction of bypass roads to divert the traffic in case the main road is blocked due to landslide occurrence is noticed near Jammu-Srinagar highway near Nashri hill area. To minimise the landslide occurrence a few more measures could also be taken. Detailed geological investigations of the area with regard to gradient, stability of slope, lithology of area, ground water conditions before undertaking construction and widening of the roads, have to be taken. In blasting, dynamite of proper intensity should be used to prevent the degradation with the rock resulting in landslide incidence. Afforestation is one of the important measures. Most suitable trees for planting in sliding slopes would be those that they have largest capacity of consumption of water and highest evaporation. It is more advantageous to plant deciduous trees than conifers which have the lower evaporation. There should be mass afforestation with species like *Robina pseudo acacia (Junglee kikar)* along the highways or consolidating the surface by network roots. *Robina* can develop long and deep roots. It is advisable to grow first grasses, herbs and hedges which will partially stabilise the slope and only after that, trees should be planted, which stabilise the slope further. Otherwise, by the time the trees sufficiently become mature to stabilise the slope, there may be landslides. Afforestation programme can be successful in areas where the landslide occurrence is comparatively less.

Rock belts are used in many countries for stabilising the landslides. They have widespread application in mining areas and tunnel operations. They help to stabilise the slope face by exerting a force which compresses the joints and prevents loosening. A rock belt also serves for greater stability by functioning as a dowel, that is increasing the shear resistance along the joints. The belts are grouted in along their entire length with cement or other chemical agents. Till now such rock belts are nowhere found in the Himalayas.

Springs emerging in the landslide areas must be ensnared and diverted to outside landslide area.

Prior warning of hazards helps to minimise the damages. But unfortunately, little effort has been made in this direction. The only Institute working in the field of forecasting is the Snow and Avalanche Study Establishment situated near Bhang, three kilometres from Manali which was set up in 1969. Nearly 80 per cent of the predictions made by this Institute proved to be true. The warning reaches the public through radio, television and newspapers mostly, at least 24 hours before the event occurs.

Although landslide occurrence which is the most disastrous ecological hazard in mountainous areas it cannot be completely prevented but its frequency, magnitude and impact can be minimised by adapting several carefully planning measures as already mentioned.

12

SUSTAINABLE LAND AND WATER USE

Of late there has been much awareness in both developed and developing countries for environmental preservation and eco-system development. With rapid growth of population especially in developing countries like India and with technological advancement in developed countries in particular, there is so much awareness of environmental degradation and ecological imbalance. Consequently, land, air, water have been polluted and various eco-systems have been affected due to over exploitation of resources of land and water with innumerable human activities like unscientific cultivation practices like shifting cultivation, industries, transport, construction activities, and over grazing by animals, etc. The land, air and water have been polluted with industrial wastes, industrial effluents, urban sewage and garbage, industrial pollutant gases like sulphur dioxide, nitrous oxide and carbon dioxide in the atmosphere by burning fossil fuels which are causing the rise in temperature resulting in rise in sea level. Besides these, there is increase of certain chemical substances like chlorofluoro carbons, causing ozone depletion in the atmosphere.

In order to protect environment from degradation, it is necessary to plan our interactions with nature and regularise our impact on nature by sustainable use of land and water resources. According to Mr G. Venkataramana the term sustainability denotes the characteristics of a process or state that can be maintained indefinitely: Sustainable land and water use is that kind of use which is adopted for protecting land and water bodies from environmental degradation and thus preserves various ecosystems of different landuse categories like forest, crop land, pasture and grazing land, etc., and water bodies like rivers, lakes and coastal marine waters, etc., and maintains ecological balance within and between them. In sustainable land and water use the emphasis is on 'improving the quality of human life within the carrying capacity of supporting ecosystems' as per the definition of IUCN, UNEP and WWF but not on maximising the outputs by overexploitation of resources.

In order to suggest management strategies for sustainable land and water use for the existing environmentally degraded areas and areas likely to be degraded in future by human activities and natural agents, they have to be identified first and then analyse the causes for such degradation in the light of land characteristics like relief, slope, climatic characteristics like rainfall, water characteristics like surface flow regimes, groundwater potentiality,water table, water quality, etc., socio-economic characteristics like cultivation practices, industrialisation, urbanisation, construction, transport, population pressure, and social/religious customs and practices which predominantly influence. The following environmentally degraded areas due to unsustainable use by human and animal activities and natural agents are to be identified in the existing landuse pattern.

I. Unsustainably Used/Environmentally Degraded Areas on Land

(i) Over exploitation of groundwater in semiarid areas for growing crops which need high water intake leading to low productivity of crops due to alkalinity of the soils.

(ii) Irrigated areas, causing waterlogging and salinity due to poor drainage.

(iii) Heavy chemical inputs for intensive crop cultivation affecting soil degradation.

(iv) Use of pesticides causing toxicity in soils.

(v) Wrong cultivation practices like cultivation along slopes leading to soil leaching.

(vi) Reduction of crop land by encroachment due to urban

expansion through construction, industry, transport activities.

(vii) Flood affected areas and wind eroded areas.

(viii) Abandoned mining areas.

2. *Unsustainably used/environmentally degraded areas affecting forest ecosystem*

Gully, ravine and bad lands formed due to floods and wind action, causing soil erosion as a result of:

(i) Shifting cultivation.

(ii) Overgrazing by animals.

(iii) Fuelwood extraction.

(iv) Timber exploitation for construction.

3. *Unsustainably used/environmentally degraded areas affecting coastal ecosystem.*

(i) Construction for human habitation.

(ii) Industrial wastes, industrial effluents, domestic sewage and garbage.

4. *Unsustainably used/environmentally degraded areas affecting urban ecosystem.*

(i) Groundwater polluted areas.

(ii) Degraded areas due to industrial effluents and hazard wastes.

(iii) Degraded areas due to urban sewage and garbage.

(iv) Air polluted areas.

(v) Degraded areas due to removal of green belts.

(vi) Degraded water bodies (lakes, tanks and streams).

(vii) Degraded areas due to construction activity.

5. *Other degraded areas*

(i) Marshy/Swampy areas.

(ii) Salt affected areas.

(iii) Barren rocky out crops.

(iv) Sandy areas.

(v) Undulating uplands.

II. Water Bodies.

(i) Inland and coastal water polluted areas due to industrial wastes, industrial effluents, urban sewage and garbage, night soil, etc.

(ii) Siltation of lakes, river beds caused by swelling and floods.

(iii) Flow affected areas in rivers through human interference by throwing garbage, forming bunds, forming human settlements and cultivation on flood plains and other landuse practices.

(iv) Ground water polluted areas due to industrial effluents, industrial wastes, garbage, sewage, etc.

The above areas are identified on the land use maps prepared for both kharif and rabi seasons with the help of satellite imageries, supplementary maps, data procured from other sources and by undertaking intensive field work to find out ground truths and thereafter analysing the causes for such degradation several suggestions like change in the cropping pattern, for example, to suggest a crop in the light of soil fertility and water availability, agro-forestry, social forestry for energy plantations, fodder crop culture, arresting constructions to prevent groundwater depletion, preventing garbage dumping, industrial waste accumulations, industrial effluents, and urban sewage to safeguard groundwater pollution, increasing green belt coverage, checking population growth and growth of slums in urban areas are made for incorporating changes in the existing land and water use pattern and finally sustainable land and water use maps are prepared.

13

DRINKING WATER CRISIS: SOME REMEDIAL MEASURES

Water is considered to be the elixir of life. A man can survive without food for a few days but not without water. Everyone knowns that water is essential not only for vital functions of human body but also for sustenance of the entire biosphere. Besides, it is needed for several man-created activities like irrigation, industries, etc. According to Dr. Rama the land surface received an estimated quantity of 400 million hectare metres (MHM) of water per year, out of which 230 MHM goes back to the atmosphere. Out of this 170 MHM, 60 MHM, sinks into ground and forms groundwater potential. Scarcity of drinking water is the most important problem in all developing countries in both urban and rural areas in the world. Most of Africa, much of Middle East, North Asia, North Western Mexico, the Western USA, part of Chile and Argentina and nearly the entire Australia are severe water deficit areas. It is more of a serious problem in developing countries of Asia, Africa and Latin America when campared to Western USA. Globally, people are suffering from water shortage. In our country more than one lakh villages are facing serious water shortage during summer. Unfortunately the significance of drinking water is not so much realised and less attention is paid to it when compared to irrigated water.

Acute water scarcity in cities of developing countries is mostly due to rapid unsustainable urbanisation. Unsustainable urbanisation never maintains the ecological balance between the urban environment created by man with his secondary and tertiary activities like industry, trade, transport, construction, etc. and the physical/abiolic environment consisting of relief soils and land forms water bodies, etc. and biotic environment consisting of plants and animals in an urban settlement. This eventually leads to environmental degradation. Here nature is more humanised than man being more naturalised and there is no harmony between humanisation of nature and naturalisation of man. This ecological imbalance makes urbanisation unsustainable. In unsustainable urbanisation the dominance of urban environment created by man over the physical and biotic environments is significant: Urbanisation is becoming unsustainable due to lack of ecological approach in urban planning. To achieve sustainable urbanisation the vertical and horizontal growth of an urban settlement should be determined on the basis of certain environmental parameters like drinking water availability, green belt coverage and cabon dioxide concentration in the atmosphere. Based on these parameters the optimum carrying capacity with reference to population should be determined and to that limit only urban expansion should take place.

The consequences of unsustainable urbanisation with reference to water resource are its scarcity, and pollution. One of the causes for scarcity of drinking water in cities and towns is over exploitation of surface and groundwater sources. This is causing scarcity of water, lowering of water table, subsidence of land, and encroaching of sea water into coastal aquifers. Lack of sufficient green belt coverage and misuse of water are also contributing significantly for scarcity of drinking water.

Over exploitation of groundwater has lowered the water table in several cities of developing countries. It has also caused subsidence of land in several cities. In Bangkok the land is sinking at the rate of 14 cms per year. At the present rate of subsidence, the whole Bangkok would be below sea level in 20 years. Similar subsidence of land is also noticed in Tokyo wherein 60 sq km of land has gone below sea level. Mexico is another city which is sinking due to over exploitation of ground water. It has sunk by 9 metres during the last 100 years. It is sinking at the rate of 40 to 60 cms per year now. In the early part of 20th century the rate of sinking was only 5 cms per year. The sinking caused widespread damage to buildings and other structures. Worst affected, are the underground sewage lines, sub-way tunnels and potable water pipes. Sinking of soil is the worst environmental problem.

There are other cities like Houston, London, Winnepeg which have undergone subsidence due to abstraction of water and lowering of water table which eventually caused subsidence.

In cities and towns located in semi-arid monsoon tropics of developing countries where the rainfall variability is high the multistoreyed buildings which accommodate large number of people, are responsible for depletion of groundwater. If the present rate of groundwater tapping continues, most of it would be exhausted in a short time and it may lead to emigration of urban people who become environmental refugees. As mentioned already due to over exploitation of groundwater, water table lowers and consequently sea water enters aquifers and the water becomes salty, as happened in several aquifers in coastal cities. As stated earlier absence of sufficient green belt coverage is another reason for inadequate recharge of groundwater aquifers.

Pollution of water is another environmental hazard of water resources besides scarcity. Both surface and groundwater is polluted largely by urban garbage, untreated sewage, industrial effluents, industrial wastes, human excreta and others. River and canal bunds in several cities are used as public toilets. Among the rivers, Ganga is the worst polluted river. Here half burnt corpses, waste materials from rituals and industrial effluents are the chief contributors of pollution to this river. River Yamuna in Delhi is another example. All the major river systems of our country have become polluted. Only 7 per cent of towns and cities in our country treat their sewage. Outside India, Nile is the worst river. Nitrates in European rivers are 45 times higher than the normal level. As mentioned already, certain religious, social customs and practices like throwing half burnt corpse in Ganga, throwing garbage, waste material of rituals, immersion of idols like Ganesh, washing clothes, and cattle are also polluting the rivers, lakes and canals.

Urban wetlands the lifeline of Indian cities for centuries protected by the tradition and preserved by people are being ravaged by sheer neglect. These lands are reduced to filthy fowl smelling cess pools, infested by mosquito larvae. Cities enjoying perennial river sources and high rainfall serve as flood cushions as in the case of Calcutta. Urban wetlands are being systematically destroyed in the name of development.

Groundwater is equally polluted all over the world by industrial wastes, industrial effluents, garbage dumps, night soil through process of percolation of rainwater and sewage. Denvor city in USA is a good example in this respect. A survey conducted in Hyderabad revealed that 60 per cent of wells, 40 per cent of borewells are not potable because of chemical pollutants like nitrates particularly in Nacharam and nuclear fuel industrial area.

Urbanisation policy makers have no comprehensive ideas regarding inappropriate pollution practices which have direct consequences on the water scarcity and pollution of groundwater aquifers, rivers and tanks. A thorough understanding between scarcity and land degradation is often lacking among technical experts.

In India, percolation of rainwater is less as compared to temperate countries like France and Germany, due to high intensity of rainfall occuring in few months. Since the period of retention of rainfall is less due to absence of vegetation cover the groundwater recharge becomes less.

One of the important measures to be adapted is, checking large-scale migration of people from small cities, towns and villages. As stated earlier, the optimum carrying capacity of population of a city has to be determined with reference to certain environmental parameters. Decentralisation of industries and other activities to small cities and towns is necessary to check large-scale migration into big cities. Storage of drinking water in percolation tanks in every city in different localities is another remedy to counter water scarcity. Tanks are the best option for rainfall storage during the monsoon months since 75 to 80 per cent of rainfall occurs in three months. The desilting of these tanks should take place every year, a practice followed in Tamil Nadu. In every village also there should be one or two well maintained percolation tanks. Community-built water tanks are found in several villages of Phillippines.

There is need to follow old and traditional methods of rainfall harvesting in individual houses. Such facilities have to be provided in existing houses and in new constructions. These methods are still followed in some parts of our country to meet the drinking water needs. In olden days houses were constructed in such a way to collect rainfall water that fell on the house. Such houses exist even today in Kutch. Earlier houses had their own wells dug below the cellar of the house. All the water that fell on the roof top was collected in the well. The water thus collected, was used throughout the year. Gandhiji's house in Porbandar has this kind of well below the cellar. People of Nagaland and Mizoram have traditional methods of water conservation. The tribal societies are unique in taking every care to protect spring and stream water sources by allowing undistrubed growth of vegetation and banana trees which are water retention species. Bathing, washing of clothes and presence of cattle are strictly prohibited here.

Mizos unlike Nagas have adopted a method of water storage since water retention has been reduced because of deforestation and soil erosion.

Sufficient recharge of borewells is needed by forming soak pits around borewells. For natural recharge of aquifers the groundwater should be porous enough to let the water soak in low lying areas to serve as natural soak pits by holding run-off to percolate down to water table. At present the entire area around a borewell is covered by roads and pavements. Even in several individual houses the whole area is covered by cemented floors. Consequently there is no soaking of rainfall recharge. To make rain water percolate into the ground it is suggested to form a pit around the borewell for percolation of sufficient rain water during rainy season. The pit may be of 2 to 3 square metres size dug up to a depth of 2 to 3 metres removing the top layer. The pit has to be filled with permeable materials like course sand, gravel, brick bats, quarry dust, etc. Water accumulating over these pits during rainy season would percolate sufficient groundwater retention. Soak pits can be cut into water logged public places by the muncipalities after confirming the presence of equifer below the top level. Soak pits not only recharge aquifers but also provide drainage to stagnant water. Loss of groundwater due to seapage into streams should also be eliminated by lowering the water table by taking out sufficient quantity of groundwater. The rise of ground table close to the surface should be prevented since it causes water logging affecting soils and crops.

There is also need to change the lifestyle and outlook of people with regard to proper utilisation of water for their daily needs. There is a misconception among people that water is the cheapest commodity available in plenty and it can be used without any restriction.

To avoid wastage of water, proper maintenance of water and sewage systems is a must. Due to negligence and poor maintenance of these systems water sometimes gets mixed up with drainage water. Old and damaged pipes should be replaced. Some of the measures adopted at the individual and family level to avoid wastage of water are: In every house taps should be closed when water is not used. Do not leave it running while washing, bathing or brushing teeth. The tap can be opened again and again when needed. While cleaning vessels, instead of soaping and then rinsing one vessel at a time, soap all the vessels at one time and only then open the tap and quickly rinse out the vessels. If the bottom of cooking vessels have got burnt of food stuck up leave the vessels soaked for a few hours and then only scrub and rinse. This will save water and save the trouble of rubbing for hours to dislodge the burnt food. If the tap is leaking leave an empty bucket under the leaking tap until the defect is rectified. Used water need not be thrown out. It can be used to water plants. Use always a small tumbler for drinking water instead of big tumbler to avoid wastage of a large quantity of left over water after drinking a small quantity. Avoid giving large get-together parties even on small and petty occasions to prevent over consumption of water and food resources. Develop an outlook that water should be consumed economically to preserve it for future generations. Bear in mind that the earth has given enough for the needy not for the greedy. Give up false notions that restricted and economic use of water is against prestige and status. One should optimise satisfaction instead of maximising output for over exploitation of resources for ever increasing wants which results not only in environmental degradation but also cause inflation.

An awareness among the public regarding the need for conservation and proper utilisation of water has to be created. Besides, certain legislative measures have to be taken and should be strictly implemented to check wastage and pollution of water.

14

TRADITIONAL RAINFALL HARVESTING METHODS

Rainfall harvesting was an ancient phenomenon in several countries of the world particularly in India. Indians since early times realised that water is an ephemeral natural resource due to seasonal character of rainfall and to overcome this setback they developed several water harvesting methods for both irrigation and drinking water purpose. These traditional rainfall harvesting methods were mentioned in Puranas, *Mahabharata*, and various Vedic, buddhist and Jain texts. There are several references to canals, tanks, wells, etc. in these texts.

Storage of water in cisterns was adapted since early times. The oldest cisterns were discovered in Palestine and Greece. These cisterns used to collect rain water from roof tops of houses mostly. The earliest dam to retain water in large quantities was constructed in Jordan around 3000 BC and Egypt in 2600 BC. Several civilisations in the world built dams and contributed to human experience and knowledge in this field. The great civilisations in the valleys of Nile, Tigris, Euphrates, Indus, and Huang rivers flourished only on the basis of water management systems. Excellent water supply and sewage disposal systems were the predominant features. As in Mesopotamia protection against the annual flooding of the Indus river, irrigation to provide assured water supply and drainage of large alluvial areas were a prerequisite to the very existence of kingdoms in the valley.

The vedic hymns particulrly those in *Rigveda* mention about irrigated agriculture, river courses, dykes, water reservoirs, wells, and water lifting structures. The relationship between water systems and environment was understood even at the time of the Greek philosopher Plato.

In Kautilya's *Arthashstra* during the period of Chandragupta Maurya (321-297 BC) several terms relating to water harvesting systems were used, for example, 'Situ' for embankment or dam for storing water, 'Parivaha', for channel, 'Tataka' for tank, 'Nadyayatana' for water from a river, 'Nadinibhandayatana' for a structure dependent on a river such as dam 'Nibandhayatana', for a canal from a river dam, ad 'Khata' for a well.

Kautilya mentioned two types of 'Situ', the 'Sahodaka' where there is a natural spring or flow of water and the 'Aharodaka' which is a storage tank with water brought into it through channels. One of the expressions used by Kautilya's *Arthashastra* is 'Adhara Paravahake daro pabhogain' which means the wet fields receiving water by channels from a reservoir.

It was also mentioned in his *Arthashastra* that the land on which the tank was built was state's property. People pooled their resources to build a tank for common use. Punishment was prescribed for a person who failed to cooperate in any kind of irrigation work. Severe fines were also imposed for damage of embankments, or flooding of the lower tank by the tank constructed at a higher level. Imposition of fines for misuse of water from dams was also mentioned. There is enough evidence in the book to indicate that people knew about rainfall regimes, soil types and appropriate irrigation techniques in micro ecological contexts.

Recent archaeological excavations near Allahabad have brought to light Indian hydrological engineering dating back to the end of 1st century BC. French archaeologicsts who discovered Harappan settlements in the Afghanistan near Russian border came across several large canals in their excavations each 25-30 kms long and some 30 metres wide.

The Harappans seem to have built an efficient irrigation system. Wells are distinguishing features of Harappan civilisations. Digging wells for water was a Harappan invention. Historians are of the opinion that Harappans used well water for irrigation for winter crops like wheat and barley. The earliest evidence of irrigation is also noticed in Baluchistan. Dams built on stone rubble were formed in Baluchistan and also in Kutch. The structures formed in Baluchistan are known as *gabarbands*. The construction of these dams involved immense skill, labour and engineering. These dams were probably used to control flood water and for retaining the alluvium coming down the hills. It is likely that they functioned as reservoirs for retaining water. Gabarbands may belong to post-Harappan period.

Recent excavations in Nagarjunakonda have brought to light some traces of an ancient canal with an average width of 50 ft. and a depth of 16 ft. It is situated in the south corner of the valley which is enclosed by lofty hill ranges on all sides.

There seems to be enough evidence that the construction of a dam and development of irrigation system took place in Mauryan period only. The Junagadh inscription of king Rudradaman makes a specific mention that a dam was built by Pushpagupta, the governor of Chandragupta Maurya (about 323-300 BC) and canals were provided by one Tushaspha under Ashoka the Great (272-232 BC).

The benevolent acts of ancient rulers to the development of irrigation facilities is also provided by 'Hathingumpha' inscription (2nd century BC) which relates to the time of first Nanda, Mahapadma, the founder of Nanda Dynasty of Magadha.

To give a comprehensive account of our country's rainfall harvesting methods developed for both irrigation and drinking water purpose, our country can be divided broadly into four ecological zones, i.e. hill and mountain regions, arid and semi-arid regions, plains and flood plains, and coastal areas.

In hill and mountain regions of India people developed techniques to divert water with the help of simple engineering structures into artificial channels that would take the water directly into agricultural fields. For diverting water from bigger streams they learnt the engineering skill of constructing bigger dams. The great ancient anicut on the river Cauvery is an excellent example of such a sophisticated engineering structure. For occasionally, the channels first lead into a storage structure to utilise the water in dry periods also. Such storage structures in Ladakh are called 'Zings'. In Southern Meghalaya, those who live near the border of Bangladesh have used intricate networks of bamboo pipelines to deliver water to betel leaf plantations in rocky areas where no channels can be built. The entire system works like a modern drip irrigation system which delivers measured quantities of water straight to the roots of plants.

For drinking water purpose natural springs are harvested in those regions. All over the eastern Himalayas and north-east hill ranges people build simple bamboo pipelines to carry water from natural springs to required spots for drinking purpose.

Rainfall harvesting from roof-tops is also common in these hill and mountain regions particularly in Mizoram wherein all houses are located either on hill tops or on upper hill slopes. Water from roof-top is collected with the help of gutters made of bamboo or galvanised sheets. The water is collected in tanks made out of tin or concrete situated either on the ground or underground. A common method of storing water is to place horizontal rain gutters along the sides of sloping roof which is normally made of corrugated iron sheets. Rainwater pours into a pipe connected to the tank. Mizos who have high rise buildings build large sized underground tanks in their basement.

In arid and semi-arid regions where the water is scarce certain methods are used for irrigation and drinking water purpose, rainfed storage structures which are called tanks were built to provide water during dry season. These tanks are also fed by the diversion channels of streams. This kind of storage tank is called 'Zing' in Ladakh, 'Ahar' in South Bihar and 'Kire' in Karnataka. The water could be used in the coming dry period for irrigation, human and animal consumption. Streamfed or rainfed storage structures sometimes built in a series with overflow from one becoming run-off for the subsequent ones. Tanks in Tamil Nadu, 'Bhandharas' in Maharashtra and 'Keres' in Karnataka belong to this system. However, all tanks are not streamfed or rainfed. Many of them simply collected water running off from a catchment area to be stored for later use. But streamfed storage system was more reliable because it collected water over a large catchment area.

Rainfed storage structures which allowed runoff to stand over and moisten the fertile soil bed of the storage structure itself, which are later used for growing crops, are known as 'Khadins' in Jaisalmar district and 'Jehads' in Alwar district of Rajasthan.

For drinking purpose, groundwater harvesting structures like wells and step wells were built to tap groundwater. These are known as 'Bevdis' in Rajasthan. Wells and step wells were also built wherever possible especially below storage structures like tanks to collect clean seepage for use as drinking water. Such structures can be found in the forts of Chittor and Ranthambore of Rajasthan. In Rajasthan also roof-top is used as a catchment area to collect rain water.

In areas where land is not a limiting factor, people even developed rain water harvesting devices called 'Kundis' to obtain drinking water. Kundis are artificial wells which store run off from an artificially prepared catchment surrounding the well so that rain water that falls on the catchment rapidly runs into the well and gets stored. Kundis are common in Thar desert of Rajasthan and they testify that any land wherever it may be, can be used as a Kundi and harvest rain water.

The nomadic Maldhars of the inhospitable Kutch region of Gujarat have developed another interesting system of procuring potable water even in an area where the rain water is scarce and the groundwater is saline. They know that the density of sweet water is less than that of saline water and hence it is theoretically possible to keep the harvesting of sweet rainwater stored in a way that the sweet water continues to float on the denser saline water and thus provide the people an opportunity to live and survive in such a hostile region. This system is locally called 'Virda'.

In the flood plains of major rivers, people build inundation channels which allowed flood waters to be diverted to agricultural fields. Flood irrigation system of West Bengal is a good example. The flood waters are also used for fertilising their fields to control diseases like malaria, by making use of fish in the flood waters to eat away the mosquito larvae. The flood plains of West Bengal developed an excellent mechanism for harvesting the rich flood waters in a measured and controlled way. In specific type of soil and cropping regions people also store rain water in the agricultural fields by bunding them. The Haveli system of Madhya Pradesh is a good example. For drinking water in plains and flood plain areas dug wells are used mostly.

In coastal areas, regulatory systems have been developed to control ingress of saline river waters especially during coastal tides and thus maintain crop productivity. The Khazana lands of Goa area are good example.

Traditional methods of rainfall harvesting should be adapted in the present crisis of water shortage. Let us assume that the average annual rainfall of our country is 1000 mm. The total population of India is 1000 million. According to a study conducted by the Centre for Science and Environment, New Delhi, the annual demand for a household water use is 34,675 billion litres and each hectare can yield 11 million litres of rain water. Thus, if India harvested rain water in over 3.15 million hectares (mha) of land or 1 per cent of India's total land area, every household can meet the water needs. Even if the collection efficiency is only 50 per cent India would need only 6.3 million hectares of land. But with urbanisation and greater areas coming under roofs and concrete structures, the collection efficiency should theoretically increase. These figures show why ancient Indians considered water harvesting structures so important and useful. it is possible that rain water harvesting structures can provide every household in India with all the water needs.

Unfortunately, the British who ruled our country, did not encourage our traditional water harvesting systems and completely neglected and destroyed them. Even the post-independence political leadership did not care to show any interest to understand these systems to serve us. The modern methods of irrigation and drinking water supply are also responsible for the decline of these traditional rain water harvesting systems.

The water crisis which is going to become acute in the coming years could be solved only by reviving these traditional systems on a war footing.

15

COMBATING FLOODS AND FLOOD PLAIN MANAGEMENT

Flood cause much damage to life and property in several countries of the world including India. In our country Himalayan rivers which are perennial cause floods. During monsoon months Himalayas receive heavy rainfall and the rivers discharge the maximum amount of water causing floods. Brahmaputra like Ho Yang Ho in China is considered to be sorrow of India in respect of flood damage. Gange and the Jamuna also cause severe floods. Even some of the peninsular rivers even though they are non-perennial, also subjected to floods but not comparable to Himalayan rivers.

Due to excess water, when the river overflows and submerges the adjacent area even sometimes causing breaches along its embankments, it is called flood. The adjoining area of the river is called flood plain. A steam overflows its normal channel and invades low levelled areas on its flood plain. The overflow occurs when the volume of water entering a stream channel exceeds the hydraulic capacity of its channel.

Floods vary in size, area inundated, duration and frequency depending on natural and certain man-made activities. The natural conditions include the quantity, intensity and geographical distributions of rainfall and snow melt, storm pattern, antecedent moisture conditions, temperature and season of the year as well as physical features of the watershed such as topography, soil, geology and drainage pattern. The man-made conditions include the various rural and urban land use practices, storage, diversion and regulation of stream flows etc.

All along the streams the flood plain may include the full width of stream valleys. In any area where floods occur the rich fertile soils deposited by such over flows and repleshment of soil moisture have sutained an abundant agriculture. Occupancy of flood plain can cause two kinds of flood losses. An individual who uses the flood pain for any development not only invites trouble for himself but also affects others by causing more flood through obstruction of its normal flood flow and reduction of flood plain storage capacity. The increased flood would affect not only himself and his neighbours but also affect residents outside the area which would not otherwise normally be flooded. The development on hillside slopes through cultivation and construction activities results in accelerated run off discharged to flood areas down stream. The periodical inundation of flood plain lands may cause loss of life, property, health hazards and affect all services to the occupants. These losses may be caused by the comulative effect of obstructions in flood plains by land use practices that cause increased flood heights and velocities and inadequate measures in raising river bunds. Any development on the flood plain must be carefully managed to minimise impacts and maximise the general benefits. To protect the flood plains proper management is necessary.

Flood plain occupancy and use is based on the economic advantages of level ground, fertile soils, ease of access, and available water supplies, despite flood risk. Full flood control is not possible or economically feasible. Abandonment of flood plain may not be reasonable because it is, not only a valuable resource but also in many areas it is the only means to support life. The nature and extent of flood plain use should be compatible with the risk involved and the degree of protection that would be practical.

Flood plain management requires a coordinated effort by water resource engineers, land use planners, economists, foresters recration specialists, environmentalists, and others concerned with an over all comprehensive approach to flood plain problems.

The flood plain management policy broadly should have two objectives: firstly to deal with immediate consequences of severe floods in a populated area, the loss of life, the physical hardship and the distruction of an environment made more fragile through human use; secondly to achieve a long-term plan for the optimum economical use of flood plains with due regard for environmental impact and socioeconomic concerns. The planners should bear in mind these two objectives for effective implementation of the management of flood plains.

A comprehensive method of flood plain management would include determining the cause, frequency and extent of flooding, determining the effect of public and private works on flood levels, preparing relevant structural and non-structural proposals for flood alleviation and determining the areas that would be proclaimed flood-prone. Comprehensive flood plain management would be most beneficial at the earliest development of a flood plain.

The alleviation or reduction of flood loss has been achieved traditionally through construction of protective works such as dams, levees, channel projection improvements and sea walls. These structural measures have partially reduced hazards and provided substantial protection in many areas. However structural solutions alone can have a negative impact on the environment.

Policies on flood plain management are now reaching a balance between structural and regulatory controls. Flood plain management including land use regeneration can affectively reduce flood losses. Landuse regulations can be used alone or in combination with protective works. Information on the extent of the flood hazard, landuse and development is essential for effective application of regulatory controls.

A well conceived and implemented land use can lead to significant reduction in flood losses with relatively small public cost and little negative environmental impact, particularly in areas where little or no development of the flood plain has occured.

Even flood prone areas can be economically utilized by suggesting proper landuse pattern.

The flood plain is an environmental and ecological resource which needs adequate attention. Flood prone areas often provide wild life habitat, water recharge areas, fertile soils, scenic areas and lands suitable for recreation and park uses. Thus the land use planner must evaluate the proper land use in terms of both the potential flood risk and the beneficial use of the natural attributes of the flood plain.

Flood plain management is most effective if a comprehensive plan is developed and implemented. This plan could consider the socioeconomic as well as environmental impacts of flood loss reduction and preventive measures.

16

FORECASTING OF INDIAN MONSOONS: THE EL NINO FACTOR

Forecasting the onset and strength of monsoons indicating whether we get normal rainfall or not is being carried out by Indian Meteorological Department (IMD) in the month of May every year. The Indian Meteorological department under the Chairmanship of reknowned scientist Dr. Vasant Gowrikar started predicting characteristics of Indian monsoon for the first time in 1987 with the help of Parametric power regression model by taking 16 parameters related to atmosphere and ocean. The I.M.D. model has been producing accurate results since 1988. In the year 1997 it was predicted that 92 per cent of normal rainfall which is considered to be normal since 9 out of 16 parameters were favourable for a good monsoon.

One of the indicators to be taken for predicting rainfall is the pressure conditions in South Pacific particularly in south-eastern part which have some bearing on the strength of Indian monsoon system whether it is weak or strong.

The pressure conditions in South Pacific ocean are somewhat related to ocean circulation giving rise to warm and cold currents on western and eastern side of South Pacific ocean. On the western side of South Pacific ocean along the east coast of Australia it is a warm current moving from north towards south and on the south-eastern side of South Pacific ocean along west coast of South America it moves from south to north where it is called cold Peru current. As a result of these cold and warm currents the sea surface temperatures vary on both sides thus modifying the pressure conditions. Since there is a cold current existing in the south east Pacific ocean along the coast of Peru and Equador, the sea surface temperatures are lower giving rise to high pressure conditions. On the other hand the pressure conditions on the West Pacific ocean along the coasts of Indonesia and Australia are comparatively lower since the sea surface temperature are higher due to influence of warm current existing there.

This pressure distribution pattern with decreasing pressure gradient towards western Pacific ocean is ideal for strengthening the south-east trade winds which cross the Equator and become strong. Deflected south-east trade winds which are known as South West monsoons, give abundant rainfall which may be normal or above normal.

But the stability of pressure distribution pattern in south Pacific is frequently disturbed in summer of the Southern hemisphere with the change in temperature conditions particularly in December onwards low pressure conditions in south-east Pacific along the coast of Peru and Equador occur due to formation of a weak warm coastal current moving southwards causing rise to sea surface temperature and consequently lowering the pressure. This warm coastal current which reverses the conditions is called El Nino or Christ's child since it occurs around Christmas time in December and coincides with southern Hemisphere summer, the season of weak south-east trade winds. El Nino may continue for four to five months. This kind of pressure variation with the high pressure around Indonesia in West Pacific ocean and low pressure in south-east Pacific ocean around Peru and Equador is called southern oscillation. With the formation of El Nino the whole phenomenon is called Enso.

During the occurrence of El Nino which rises the sea surface temperature the fish stock disappears and is not available to fisherman and the coastal birds which depend on fish for food die in large numbers. Normal arid areas of northern Peru and Equador suffer under torrential rains. On the other hand dry conditions with droughts and low rainfall may be caused in Indonesia, East Australia and India.

The weak pressure conditions arising out of the formation of El Nino cannot strengthen and speak up the south-east trade winds as such they become weaker. With the result a weak monsoon system develops and low rainfall or droughts may occur in Indonesia, Eastern Australia as well as India.

Many foreign scientists particularly from USA have predicted that there would be adverse impact of El Nino on Indian monsoon system giving low rainfall. A German newspaper *Oil Wordd* has also predicted the occurrence of droughts in south-east India and Australia with the formation of El Nino.

But the Deputy Director General of Indian Meteorological Department Dr AVR Krishna Rao stated that since El Nino is one of the 16 parameters taken into account its effect would not be much. On account of nine favourable factors out of 16 this year, 92 per cent of normal rainfall was predicted. The forecast was made on the basis of Power regression model which was developed with the experience in forecasting rainfall over the past 100 years.

The adverse impact of El Nino on Indian monsoon system may not be hundred per cent accurate since on eight earlier occasions when the El Nino factor was unfavourable we received below normal rainfall in four years only. Since El Nino is one of the factors out of sixteen taken into account, its adverse effect may be minimised or compensated by other favourable factors. This year 9 out of 16 parameters are favourable and moreover the snow related parameter which has a strong influence on our monsoon system is good this year.

17

AGROFORESTRY FOR STABLE ECOSYSTEM

Natural resources like forests, agricultural crops and livestock are very essential for the development of any country. They meet several necessities of man for his living. Among these, forests have wider uses. They not only check soil erosion and increase rainfall, but also provide sufficient oxygen for the survival of man and prevent environmental pollution. Besides they provide wood for construction of houses and bamboos as raw material for paper industries. Minor forest produce serves as raw material in several industries and in the preparation of drugs. In spite of innumerable benefits we get from forests, they are destroyed for extension of cultivation of land, construction of houses, and extension of transport facilities and also due to over grazing by animals. Every year on an average 15 million hectares of forest area is destroyed. Much of this destruction is caused in Asia, Latin America and Africa. Many scientists are of the opinion that in about 20 to 30 years about 10 lakh plant species may completely perish. It is estimated that about 10 million tons of carbon dioxide is added to the atmosphere due to the destruction of forests every year.

In our country destruction of forests is taking place on a large-scale particularly in the Himalayan region. According to National forest policy 60 per cent of the area should be covered by forests in hilly regions. But in Nagaland which is predominantly a hilly state, the percentage of forest area is slightly over 17 per cent. Here deforestation is taking place due to shifting cultivation by burning forests and for providing wood pulp for paper industry. In our country the percentage of forest area is 22.86 only instead of 1/3rd of total geographical area as per National Forest Policy.

Due to destruction of forests, not only the forest ecosystem is affected, it is causing large-scale environmental degradation. To prevent this degradation, a new plantation system called agro-forestry is being adopted now in several areas and much publicity is given to it. No doubt agro-forestry can be an environmentally sound and economically feasible means at developing and protecting marginal and economically fragile lands.

Agro-forestry is a broad term for any landuse system that combines, trees, crops and or animals in an interactive manner either simultaneously on sequentially on the same unit of land.

Agro-forestry system involves atleast one species of woody perennial with atleast one species of plant or animal. Agro-forestry systems produce atleast two outputs and have a cycle of atleast one year.

Agro-forestry landuse system is highly complex which needs a multidisciplinary approach and it should be undertaken with adequate understanding of cultural, economic and environmental interactions.

The various forms of agroforestry can be broadly classified into three categories: (i) agro sericulture, wherein agricultural crops are mixed with woody perennials; (ii) silvo-pastorolism, wherein livestock is integrated into forests or tree crop plantations; and (iii) agro-silvo pastoralism wherein agricultural crops, woody perennials and livesock are integrated. Agro-forestry systems have evolved where either forestry or agriculture were the major landuse. These systems are especially important in areas where it is vital for poor, low input farmers to produce variety of products for their sustenance since they cannot depend upon one system only due to unassured and inadequate outputs.

The basic objective of agroforestry is to increase, diversify and sustain farming output in a manner that is both ecologically and socio-economically feasible. The positive interactions between environment and agroforest components could maximise land productivity and sustainability. As such it can be called sustainable landuse system. Agroforestry can serve to reduce the degradation of natural forests by providing alternate sources of food, and by making forests an integral part of local socio-economic system. The major ecological potential of agroforestry exists in marginal, low input farming areas where woody perennials can contribute to soil fertility and protection. The woody perennials can provide atmospheric nitrogen, carbon, air, etc. Soil through root action, prevents: (i) soil erosion and subsequent fertility loss by providing vegetation cover; (ii) modifies soil porosity and unfiltration rates; (iii) enhances nutrient cycling economy; (iv) stabilises soil temperature, resulting in more balanced microclimate; (v) increases permeability and waterlogging capacity and trap with deep roots, nutrients otherwise being lost. Trees also positively affect the broader environment by elimentation of the water table and natural overflow channels by internal drainage of excess waters, and by reducing in rate siltation in dams and reservoirs. There are also other beneficial effects on the air quality, radiation levels, temperature, humidity and rainfall.

The choice of species in agroforestry depends to a large extent on the ecological conditions of the site in question. Planting species unsuited to the local ecology is not economically cost effective and can lead to serious environmental degradation. It is also important to consider the natural compatibility and non-competitiveness of the various agricultural and tree species to be introduced. The different components of an agroforestry system should be complimentary over all stages of growth and maturity.

In cases where forestry is the dominant landuse, the choice of agro crops is decided by which species can grow under prevailing conditions without posing a threat to woody perennials. In cases where farming is the dominant landuse the agricultural component of agroforestry is largely influenced, by local socio-economic factors, the selection of species rests mainly with woody component. Special regard should be given to avoid species that pose problems of competition of water, nutrients, and solar energy.

There are three types of plantations in agro-forestry. They are mixed cropping, zonal cropping and rotational cropping. Under zonal agro forestry, the species are arranged in zones (strips, bands or other geometric shapes) so that there is some but not excessive interaction between species. The decision to use any one of the three methods depends primarily on the degree of compatibility of species in terms of resource use and ecological setting. Where species are highly complimentary mixed cropping is viable. If the biological interaction is not so favourable some form of zoning should be considered. The exact configuration of zones depends of the degree of positive interaction. Special consideration should be given in planting trees in a manner suited to soil and water conservation. Planting the wrong tree in a wrong place can lead to disastrous ecological effects.

The choice of agroforestry is strongly influenced by soil, climate and topography. For example, in dry and semi-arid regions where there is water scarcity, trees which consume less water, and food crops like millets and grasses as fodder for livestock can be grown. Trees are planted against the direction of the winds to prevent soil erosion in these areas. Species such as Acacia Albida, and Prospis cineria which consume less water are recommended for sandy, low rainfall areas, but the choice of tree species must be determined locally.

In humid lowlands the agroforestry system is quite different. These areas get high rainfall and are subjected to intense farming. In these areas hedges are planted with fast growing woody perennials and plantations for fuel wood production and trees which can withstand floods are grown. Crope like rice and jute which can withstand heavy rainfall and need lot of moisture during growing season are cultivated. Trees like *Sasbima grandiflora* which has tolerance for flooding and which requires little maintenance are grown in these areas.

In tropical mountains and upland regions the agroforestry ecosystem differs from others. In these regions soil erosion and uneven topography are the major problems. Such trees with soil binding capacity for preventing soil erosion, and for providing food, fuel, and fodder are grown in these regions. The trees grown here are multipurpose type and are planted along contours. The agricultural crops are grown between rows of trees, planting of soil binding grasses takes place between the trees along contours. Examples of high altitude species include *Aclanthms altisimma* which can grow in a variety of acid and alkaline soils and can tolerate flooding and drought.

As mentioned earlier, the types of trees, crops and grasses

grown under agroforestry system not only depends upon physical characteristics of an area but also depends upon the existing landuse pattern. Woody perennials can be introduced into farming areas, crops and livestock into tree crop plantations and forest plantations, and woody perennials into pasture lands.

In farming areas where intensive cultivation exists, lot of sunlight is required and as such farmers are reluctant to plant trees since they cover the fields with shade. Moreover the farmers concentrate more in cultivating high yield crops as the possibilities of planting trees are limited. But trees like coconut and mango could be grown on unused high level plots of land, along canal bunds and roadsides. They can be grown in fields of such crops which can withstand the shade of trees. Similarly grasses which can withstand shade of trees can also be grown in fields. Shade tolerant crops like coffee, papaya between the trees can be cultivated. To increase soil fertility leguminous crops are grown. If fruit crops are grown in forests, tribals who cause large-scale soil erosion by practicing shifting cultivation for their livelihood could be diverted from this practice to horticulture as their occupation to earn their livelihood. Our government is keen in growing fruit crops in forests.

Thus, agroforestry is an attractive means of combating the ecological and socio-economic problems of the tropics, by improving land productivity, protecting soils, and providing wood, food, and cash crops in a manner acceptable to farmers it makes it possible to give a self-sustaining and stable ecosystem.

18

PRESERVATION OF COASTAL ECOSYSTEM

Ecosystem comprises the interactions and interrelationships between and within the non-living i.e., abiotic environment consisting relief, soil, climate, water while the living, to, biotic environment consists of plants, animals including man in a spatial unit of the earth. In other words, the study of an ecosystem is primarily concerned with the structure (composition) and functioning of living system with reference to a non-living system. Depending upon the area of operation of various elements belonging to biotic and abiotic environments, different ecosystems have come into existence each with its own distinct characteristics. For example, a coastal ecosystem is different from a lake ecosystem and so on. In an ideal ecosystem an equilibrium or stability is maintained between these two environments, which is called ecological balance. Often this ecological balance is disturbed due to over interaction and dominance of biotic community particularly man over abiotic environment, who exploits nature to satisfy his wants and desires, causing environmental degradation.

In fact the damage caused to environment by several abiotic physical agents like earthquakes, volcanoes, floods, storms, etc., is much less when compared to that of man.

This study deals with coastal ecosystem which is one of the most important ecosystems, as such its protection is vital for the very survival of man.

While the shoreline is defined as the water's edge at various stages of tide, the coastline is mentioned as the high spring tide shore line. All the area lying a few kilometres away from the shore line upto a few kilometres below the shore line is generally considered to be the zone of coastal ecosystem. The coastal area may be formed partly as a pereplane due to ercosional process or by process of deposition by waves and partly by the uplift of the continental shelf. An emerged coastline due to uplift will be straight like the east coast of India or indented due to submergence like the coast of Norway. The coastline is subjected to the process of degradation by waves and the eroded material is deposited along as well as off the coast. As such a coastline is always in the state of dynamic equilibrium and is characterized by a number of features like beaches, lagoons, rock islands, off-shore sand bars, spits, river deltas, etc.

Depending upon this abiotic environment of the coast, the biotic environment attains certain characteristics, giving rise to a typical socio-economic system. A typical coastal area is characterized by the growth of mangrove forests, development of fish culture, the rise of ports, human habitation particularly fishermen settlements, rise of densely populated cities and towns, rise of several industries taking advantage of port facilities and rise of star hotels for recreation and tourism in places like Goa. But man with his supremacy over biological agents with his advancement in science and technology interferes with ecosystem and disturbs the ecological balance to satisfy his unlimited wants. The destruction of ecosystem by man is several times higher than that of physical and biological agents and is threatening his very existence. Many people living along the coast are on the edge of survival.

Sea coasts throughout the world in both developed and developing countries are used are toilets and garbage dumps. Even the industrial wastes are dumped along the coasts and many industrial effluents charged with toxic chemicals pollute marine waters. Coastal resources like mangrove forests crests, coral, reefs, fish, etc., are being exploited recklessly.

The widespread degradation of the coasts is also due to over population particularly along the coasts of India. Along the coasts of the many countries population densities reached over 500 per square kilometre. On a global basis probably half of humanity or 2.5 billion people live within 100 kilometres of a sea coast.

Coastal environment degradation is also due to mismanagement of land. Huge quantities of sediment is transported into sea through rivers, due to destruction of forests in water sheds resulting in decay of tidal mangrove forests, coral reefs and fish. It has been estimated that from Bangladesh and Nepal about 1.46 billion tons of sediment is being deposited into the Bay of Bengal every year by the rivers. Run off from fertilizers and pesticides is causing much damage to the marine flora and fauna tragically the loss of these vital resources is due to ignorance and greed of the people. In many parts of the world, coastal people do not understand the role of coastal ecosystems in providing them food and other necessities. The importance of coastal wetlands as nurseries and feeding areas for fish cannot be underestimated. Globally, nearly 2/3 of all fish caught are hatched in tidal areas. About 80 per cent of Indian fish caught from the lower delta of Ganges and Brahmaputra rivers come from mangrove swamps of Sunderbans which cover 6,000 square kilometres. Pacific islanders depend upon marine resources for at least 90 per cent of their protein intake.

Mangrove forests wich are salt tolerant are considered by biologists as most productive and diverse wetlands on earth supplying important habitats for over 2,000 species of fish. The roots of mangroves provide sanctuary for sponges, crested worms, molluscs as well as green, red and brown algae, etc. Besides they act as shore stabilizers by checking erosion of land by sea waves and serve as sheltert belts to minimize the loss of life and property by cyclones and stroms. They also supply fodder, building material besides oil, wax, dyed and tannin and medicine. Besides, they serve as important wildlife habitats.

For a Breath of Fresh Air

The loss of these forests is also due to run-off charged with pesticides used in agricultural fields, floods occurring due to deforestation in unplanned water sheds siltation and mining operations, etc. The worst destruction of these forests is taking place in Asia. India lost over 16,000 sq. kms of mangrove forest area since the beginning of this century. The total mangrove forest area at present is about 6,740 sq. km. Out of which 4,200 sq. km is in Sunderbans. In Sunderbans 1,500 sq. km of forest area has been cleared for agriculture. The forests area also affected by an estimated quantity of 20,000 kgs. of pollutants per day which is being brought in and deposited in the estuary and these include some of the toxic chemicals.

Most of the mangrove forests in India are inhabited by the poor tribal communities whose livelihood depends upon the resources of these forests.

The coastal ecosystem is also characterized by the presence of a large number of coral reefs, located a few kilometres away from the coast in different parts of the world. The coral reefs cover about 6,00,000 sq. km. of area in the tropical region. Most of these reefs are important sources of forests and a variety of fish. They provide habitat for over 4,000 species of corals and different molluscs. Besides they act as barriers against storms. But coral reefs are exploited and damaged to a large extent through deforestation along coasts and uplands, coral mining and blast fishing operations.

With the destruction of forest for timber and agriculture, large quantities of eroded sediment will reach coastal waters and is often transported to coral reefs where it suffocates the living corals. Large-scale damage is also caused by blast fishing and mining converting large areas of coral reefs as deserts. The increasing use of poisons and dynamite for blasting by poor fishermen has converted coral reefs and they are in serious conditions. As mentioned already the coastal ecosystem is badly affected by industrial wastes, sewage and industrial effluents, garbage, night-soil and other filthy material contributed by human beings.

Some of the industrial wastes reaching the coastal waters are very harmful affecting flora and marine fauna especially fish and causing diseases to human beings. Methyl alcohol polluted Japan waters affecting fish and two thousand people having eaten this fish suffered from neurological disorders. Some people also got an attack of cholera and other diseases by eating fish of polluted waters. Several people living along Kerala coast are affected by radiation on account of the presence of monozite mineral in beach sands.

The rising of sea-level due to global warming has the worst impact on the coastal ecosystem. It has been estimated that by the end of the this century the temperature of sea would rise by 2 to 3°C. Global warming is mainly due to the rise in carbon dioxide (CO_2) concentration in the atmosphere due to industrialization, development of transport, rise in population and large-scale deforestation. The increase in temperature causes the rise in sea-level temperature due to thermal expansion of sea water when warmed and also due to melting of snow peaks. A rise in sea-level by 1 metre would inundate coastal lands particularly river deltas, mangrove forests and coastal erosion, threatens coastal structures increases salinity in estuaries, rivers and coastal aquifers and submerges river deltas. Many islands like Maldive island would submerge if the sea-level rises by two metres.

One metre rise in the Bay of Bengal would also flood 15 per cent of the land, killing thousands of hectares of mangrove forests in the Sunderbans displacing atleast 10 million people. It has been also estimated that 40 per cent of Bangladesh would be submerged by the year 2000. The most affected would be Lakshadweep islands which are lying not more than few metres above mean sea-level.

In order to preserve coastal ecosystem the coastal zone management programmes have to be undertaken vigorously by involving local communities from the cities down to the fishing villages and who have to be given a strategic role in conserving resources and carrying out these programmes.

One of the most important measures to be taken under the coastal management programmes for preserving coastal ecosystem is to prevent any built up structures below the 500 metres High Tide Line (HTL) of the coast. Unfortunately, in many countries particularly in India for the development of tourism, the Tourist Development Corporation of India has encouraged a number of star hotels to come up along the coasts particularly along the coast of Goa. Many of these hotel managements violated the environmental norms and extended their structures even below the 200 metres high tide line. Now the environment ministry has taken up this issue seriously and asked the managements to demolish all structures existing below 500 HTL.

The high concentration of population along the coasts due to large-scale migration from the land areas should be checked by decentralization of many activities particularly shifting of industries. The high density of population is not only polluting waters by contributing garbage and by over consumption of coastal resources but also causing atmospheric pollution by increasing carbon dioxide content which is contributory green house effect gas for raising the global temperature and consequently rise in sea-level which submerged vast extent of coastal areas. The increase of green house gases particularly CO_2 is not only increasing of heat caused by the growing population but also by the indiscriminate consumption of fossil fuels by vehicles used by the people for transport. As a result many coastal towns and cities are warming up rapidly. To preserve coastal ecosystem the optimum size of a city or town with reference to pollution should be decided on the basis of environmental parameters like availability of coastal resources particularly the extent of barren and uncultivable land for built up area and the land available for parks and gardens, etc., to maintain the same concentration of CO_2 in the atmosphere. In other words, the optimum carrying capacity of a city or town with reference to population should be decided keeping in view the CO_2 contributed by the number of people in relation to vacant land available to support vegetation, form absorbing the CO_2 contributed by the people, in oder to maintain the ecological balance. To maintain this ecological balance the migration of people from inland areas to coastal areas should be checked. Any increase in population needs additional increase in vegetation cover to maintain the normal levels of CO_2 concentration in the atmosphere. In other words, the optimum carrying capacity of population by the coastal region should be decided to keep the balance between the CO_2 contributed by the people and the amount of vacant land available to support vegetation to absorb the CO_2 emitted by the people. To maintain this ecological balance the population migration to coastal areas should be checked.

Decentralization of industries and other activities from cities and towns would check the growth of slum dwellers who have a major share in causing air and water pollution. As mentioned already checking soil erosion, floods and siltation in the coastal waters by taking up afforestation programme use of organic manures in place of chemical fertilizers, abandoning the use of pesticides to make the drained water from agricultural fields free from pollution, using the urban garbage for generating power, stoppage of dumping hazard wastes along the coasts are other important measures to be taken up for the preservation of coastal ecosystem. Over and above change in outlook and life-style in the people is most essential.

Any programme for environmental preservation can be successful if people are educated about the need to preserve environment by creating an environmental awareness among them and also physically involving them in its implementation.

Coastline Protection From Degradation

The coastlines of all oceans is subjected to constant erosion by waves, tides and as such they are in dynamic equilibrium. The coastline is frequently referred to as shoreline. Strictly speaking a coastline is different from a shoreline from the geomorphological point of view. The shoreline is the water's edge at various stages of the tide, whereas the coastline is the high springtide shoreline. The area lying between coastline and shoreline is called the coast. Protection measures are needed for coastline as well as area lying between the coastline and the low tide shoreline. Basic objectives of coastal protection measures are concerned with protection of coastal property and infrastructure from inundation by the sea and to retain a usable beach.

Coastal protection measures must operate within both the immediate and the broader geomorphological contexts. No part of any geomorphological or ecological system operates independently of any other.

Over the past few decades an estimated seventy per cent of the world's sandy shorelines have experienced erosion. Natural factors including sea-level rise may partially explain this phenomenon. However, man's activities have undoubtedly led to significant localized erosion. Dam building for example, has cut off a major source of sediment supply to many stretches of coast around the world causing considerable loss of land. Sand extraction has directly reduced the amount of material available to nourish neighbouring beaches and sediment supplies have been often interrupted by protruding structures such as jetties. Where these interferences are identified as the source of erosion problem, appropriate protective measures are to be undertaken.

Coastal erosion and flooding are natural phenomena on many stretches of shoreline. These phenomena would become problems when infrastructure is located in the zone exposed to coastal processes. The erosion and flooding threat to property can be countered in a number of ways. These can be broadly divided into measures which require the construction of coastline structures and those that make use of geomorphologically compatible solutions. Often more than one measure is used but a combination solution requiring the use of two or more options may in some cases be most appropriate. It might employ a structural method such as a sea-wall with an environmentally comptible method such as beach nourishment or aquatic vegetation. The weighing up of various alternatives requires careful consideration of the coasts and benefits of each protection scheme. Short and long-term physical impacts, both at the site in question and in areas down-drift are key considerations.

Coastal Protection Structures

Coastal protection structures are commonly employed to safeguard coastal developments in the event of storm surge or to remedy the erosion that results from unfavourable rates of longshore drift. Coastal structures inspire a false sense of security which encourages higher building densities closer to the shore. The risk to life and property in the event of structure failure can therefore be greatly magnified.

Sea-Walls

Sea walls, revetments and bulk heads perform similar functions. They are used to protect the shore and pevent inland flooding on beach, cliff and estuarine coastlines. Sea-walls are usually vertical or near vertical impermeable structures made from concrete, wood or steel. Revetments are usually made of boulders which form mounds or slope armouring to prevent erosion in estuarine or lagoonal locations. Bulk heads can be made from a variety of materials from old tyres to steel pilings. They serve the same function as sea-walls. The impact of sea-wall revetments and bulk heads are to a large extent dependent on the environmental setting. Sea-walls can have serious economic and environmental consequences if they are not used in conjunction with other protection measures.

Massive sea-walls have been often constructed on the upper beach of sandy shores to protect coastal property. Even in low energy environments, such walls modify shoreline processes. The sequence of modifications lead to the loss of the protective beach and may culminate in the undermining and collapse of the structure itself. Larger and more expensive structures are then required and maintenance costs increase.

Revetments, bulk heads, dykes and levees are some of the names given to walls in estuarine or lagoon locations. Their primary purpose may be to protect low-lying developments from flooding to stop coastline erosion.

Breakwaters

Offshore breakwaters protect the threatened sections of coastline by reducing incoming wave energy. They are usually rubble mound structures and may be exposed or submerged, permeable of impermeable. Traditionally they are used for harbour protection. They are detached shore-parallel breakwaters.

Groins

Parallel timber or groins have traditionally been constructed to increase the width of the beach or reduce the rate of beach loss. These structures reduce the natural rate of longshore drift by acting as sediment trap, thereby allowing sand accumulation. Groins are placed perpendicular to the shore individually or in series and often from sea-walls to provide protection to the beach.

Geomorphologically Compatible Methods of Coastal Protection

Traditional structural methods of coastal protection have remolded stretches of coastline, beaches having been lost in the process. Flexible geomorphologically compatible methods using natural materials which are able to adapt the prevailing natural conditions are better to satisfy recreational and aesthetic requirements. Indeed geomorphologically compatible methods may help to form valuable new habitats.

Beach Nourishment

Beach nourishment with sand bringing from outside when the rate of sand removal exceeds supply to meet the deficit is one of the methods adapted. The fundamental advantage of this method is that it counteracts sediment starvation which is often the root cause of the problem. Material is dredged from areas of sediment accumulation. The borrowed material is then allowed to drift naturally from 'feeder beaches' to the beach in question or spread directly on to the eroded beach both above and below lowwater mark. The re-nourished beach is similar in appearance to the beach prior to erosion.

Beach nourishment provides a wide recreational beach and affords protection to beach property, albeit at considerable expense. The method is less technical in nature when compared to conventional structural methods and has the added advantage that it can be discontinued or followed up at any time. However, the method may cause environmental damage in sensitive locations. Unless the beach nourishment material is free from fine silt and clay particles, the impact on flora and sedimentary fauna may be adverse. If the material contains lot of fine sediment it settles over an extensive area, smothering and killing sensitive marine organisms. Coral can be harmed by the reduced light intensities and increased turbidity and killed by sedimentation. Reefs are the substrate and shelter for many sedimentary colonizing organisms which are highly susceptible to siltating and cannot reproduce quickly to recolonize the reef after disturbances. Commercial and artificial fisheries can be adversely affected, kelpbeds and shell-fish reefs are also affected by sedimentation.

Beach nourishment requirements should be met by using offshore deposits. Near shore sources of beach nourishment should not be used. Offshore continental deposits are popular sediment sources because they often contain large volumes of material of the appropriate quality.

Artificial Dunes

The construction of vegetated dunal ridge to trap and retain wind blown sand provides a natural, flexible form of protection against coastal flooding and at the same time creates a valuable habitat for nesting birds. A dune is the mobile equivalent of a sea-wall, acting as a dyke protecting inland areas against rising water-levels. It also acts as a reservoir of sand which can be transferred to the beach during periods of erosion. This movement of material enables the beach dissipate storm wave energy in an optimal manner. Dune building can be accelerated using semipermeable fences, or bulldozing sand into aerodynamically stable forums and the ridge can be held in place by planting dune grasses and shrubs. Vegetation is self-propagating and self-sustaining, so unlike traditional structural remedies, maintenance costs are minimal.

However, where coastlines are naturally retreating, dunes will gradually be destroyed. If natural materials and nature plant species are used, artificial dunes can diversify the shore ecosystem and provide a valuable habitat for a variety of birds, mammals and reptiles.

Aquatic Vegetation

Salt-marshes bordering estuaries and low energy beaches can be established to form a protective apron against sea attack. They retard erosion along unconsolidated coastlines by breaking the force of the waves and dissipating wave energy. In estuarine locations salt-marshes perform the additional role of trapping silt and gradually raising the level of the shore. Temperate and sub-tropical species of spartina—a vigorous salt-marsh grass, have been used successfully in coastal protection projects in many countries. Sea-grasses planted sub-littoral environments can substantially reduce current velocities and hence may retard longshore drift. Dense thickets of mangroves present an effective barrier to storms along low-lying tropical and sub-tropical coasts and may be planted specifically for this purpose.

Salt-marshes provide a highly productive ecosystem which may support a range of invertebrates, wildfowl and mammals. Over a period of years salt-marshes may form grazing pastures which can be used by cattle and sheep. Sea-grasses can support a plethora of marine life, such as crab, shellfish and other food animals. Both salt-marsh and sea-grasses likely to be more productive than the ecosystem they replace. The same is true of mangroves. They serve as nursery grounds to species of economic importance and provide a wealth of scientific and wildlife interest. The planting of aquatic vegetation should be encouraged, wherever suitable growing conditions exist.

The large volume of sand held by vegetated frontal dunes provide one of the most valuable froms of naturally available coastal protection. These dunes are important nesting areas for sea-birds. Efforts should be made to ensure that the binding vegetation is not lost as a result of excessive, recreational pressure, over-grazing, or burning. If the vegetation cover is removed, dune sands can be blown inland and lost to the beach and dune system, leaving coastline development vulnerable. Use of dunes as a source of sand for beach nourishment is to be avoided. The protection function is lost when they are isolated from their source of nourishment by buildings and developments and should therefore be located landward of the active dune system.

An appreciation of the value of mangrove swamps and salt-marshes as natural forms of coastal protection usually comes too late. Mangroves favour accreting coastlines. Where groins or jetties interfere with the sediment supply, mangroves may be damaged. Artificial sand by passing systems can be installed to rectify the sediment imbalance.

The same chain of events occurs along coastlines with offshore coral reefs. Reefs influence the development of tropical beaches through their natural break water function.

Mining damages reefs, reducing their effectiveness as break-waters. Storm waves can then enter the near shore zone and this leads to beach disappearance and coastal retreat. Fishermen have also played a part by widening existing breaks in the reef and creating new ones and also through dynamiting fish. Such harmful activities may present complex management problems. Solution must based on a careful evaluation of socio-economic and cultural ramifications of control measures.

The choice and design of coastal protection measures require and understanding of the processes involved in shoreline change. The shoreline is the interaction between land and sea. The most important factors govering the interaction are the land-forms occuring along the land sea interface, the materials of the coastal zone and the prevailing processes of which wind and waves are usually the dominant components.

Detailed knowledge of the local situation is also a prerequisite. Sufficient date must be acquired to take management decisions.

As such coastal protection measures need lot of careful planning after collecting and processing of data related to wind and wave conditions, mean sea-level change, beach profile variations, sediment sources, ecology, etc.

19

POLLUTION THREATENS KOLLERU LAKE

Kolleru lake in Andhra Pradesh is a scenic spot and presents a picturesque landscape. It is an important problem area drawing the attention of the planners for ecosystem development, being one of the largest lakes in India.

The lake is a shallow depression situated in the interdeltaic area of Krishna-Godavari rivers (16°–32' N to 16°–47' N latitude and 81°–5' E to 81°–21' E longitude) in the coastal plain of Andhra Pradesh. It was originally a part of the Bay of Bengal, but during the pleistocene period (between 1 million year-25,000 years BC) it was formed into a lagoon which was subsequently cut off from the sea with the coalescence of the deltaic deposits of Krishna-Godavari rivers. The lake is connected by a tidal stream known as Upputeru, 27 miles in length. The lake collects the drainage from a few streams and delta drains and has a catchment area of 2,823 sq. miles. The highest water-level is 10 ft. contour (MSL) and the lowest is 0.2 ft. contour (MSL) which is the dead storage level. The deepest level of the lake is –2ft. M.S.L. with fluctuations in water-levels the water spread area changes. The maximum water spread at the high water-level is 305 sq. miles and the water spread area is 2.6 sq. miles at the lowest water-level.

Historically, Kolleru lake was well-known and the famous Mahakavi Dandi in his work *Dasakumara Charitra* mentioned about it. According two copper inscriptions belonging to 1377 and 1382 AD obtained in Kolleru lake a big fort was existing perhaps in the present vacant site of Kolleru Kota during the regime of king Langulya Gajapati of Orissa. In the inscription of Pulakesi-II, it was mentioned as Kolanu Veedu. It was also known by other names such as 'kolanu' and 'koonalamu' in the historical past.

There are 88 villages in and around the lake and most of them are surrounded by water and look like islands, when the lake gets flooded during the rainy season. The lake has gradually become shallow due to enormous quantities of silt deposition by the streams and drains falling into it. The lake is overgrown with weeds. It is mostly inhabited by fishermen.

Until recently, the most important problem was the rapid rise of water-level during the rainy season, submerging vast areas of cultivable lands and causing considerable hardship to the inhabitants of the region. The rapid swelling of the lake was mainly due to low discharge capacity of its outlet Upputeru, while the total inflow contributed by streams and drains entering into it was much higher. As a result of this great difference between inflow and outlow rates, the lake used to get inundated rapidly. Excessive growth of weeds in the lake, cross bunding the lake by cultivators and meandering course of the outlet, also contributed to the low rate discharge from the lake. To keep the lake water-level below 7 ft. contour, the government of Andhra Pradesh has taken steps to increase the discharge capacity, by widening and deepning of the outiet and also by constructing detention reservoirs on some streams to minimize inflow.

The lake is full of aquatic weeds of which water hyacinth, 'Tunga' (*cyperus articulatus lin*) and 'Jammu' (*typha elephantina roxb*) cover a large part. The last two weeds are used as roofing material located in and around Kolleru lake. The weeds prevent free flow of water from the lake and obstruct navigation.

The lake abounds in large varieties of fish, like eels, murrels, etc. Varieties of birds are also seen in the lake. Some of the birds like pelicans and cranes migrate from far off countries like Siberia in the month of October and leave in the month of April. At one time, there used to be large number of birds and the hunting of these birds had become an occupation for some people. Some very interesting bird hunting techniques were popular here in those days. But with the starting of cultivation and human habitation on elevated islands, the natural habitat of migratory birds got disturbed and the number is very much decreased. Besides birds, a large number of ducks are also reared by the people living in the peripheral villages of the region.

The aquatic environment of this region is congenial for the development of cultivation, fishing and duck rearing. All these activities shift in real extent with reference to water-levels in different seasons. While cultivation and fishing form the principal occupations of a large number of people, duck rearing is done by a few individuals living in peripheral villages only.

The cultivation in Kolleru lake has got its own characteristics, due to inundation. It involves much expenditure and requires much labour. Rice is the only crop grown in both *kharif* and *rabi* seasons. To make the lad fit for cultivation, it must be cleared of weeds. Special varieties, which can resist flood are grown. Harvesting of the crop is sometimes done in standing water using boats and canoes.

For cultivation in rabi season, which commences in January, the water has to be baled out first and stores in tanks and pumped back into the fields and this is a labourious and expensive process. However, the yields of *rabi* crop are higher due to fertility of soil which is enriched by silt.

Fishing is another important occupation. Fishing activity is highest from September to January. The lake has large varieties of fish among which carps, catfish, murrels, eels are important. Conical and rectangular basket traps, murrel nooses, are operated in the lake for catering fish. Nets cannot be operated in the heart of the lake due to weedy growth.

Palmyra canoes are mostly used by fishermen to move about in the lake for catching fish. Most of the fish both live and dead is exported to Calcutta. Due to higher income obtained from fish, when compared to cultivation most of the private people are coverting their lands into fishing tanks. An acre of fish tank may fetch an income of Rs. 10 to 15 thousand rupees and this is much higher than the income from one acre of land. Fishing is mostly done by 'Vaddis' who are of fishermen by caste and they form the bulk of the population. Fish preserving and marketing facilities are however, inadequate. Ice is brought in huge quantities from cities like Vijayawada. Recently an ice factory has come up at Akividu, the principal marketing centre, but the quantity produced is inadequate.

Duck rearing is concentrated in the peripheral villages of the lake due to space requirement and accessibility to large urban and metropolitan markets. The peripheral towns such as Eluru, Kaikalur, Tadepalligudem, Akividu are the main centres of duck rearing. Ducks are reared mostly in Kolleru tract in November-December and May after the harvest of *kharif* and *rabi* corps. Duck eggs are in great demand in West Bengal and the bulk of the export goes to Calcutta.

The demographic structure and settlement patterns are also influenced by the hydrological regime of the lake. The peripheral villages, being located on higher contours and with proximity to transport routes, show higher densities. The literacy rate among population is also higher in peripheral villages. With the approach to the interior of the lake, the settlement size shrinks. The settlements are compact in the interior of the tract and are mostly inhabited by fishermen. Villages closer to transport lines have specialized functions. As such, sole fishing or sole cultivation villages are aligned to transport routes villagers situated on lower contours have fishing and cultivation as their main functions. Some of the villages with little transport are lacking any functional specialization. House structures are mostly rectangular or circular in shape. Aquatic weeds are used as roofing material in most of the villages.

Inaccessibility is another major problem in this region. A large number of villages have no educational, medical, communication and drinking-water facilities for a number of years. However, during the last two decades there has been considerable development regarding these facilities in the region. A large number of villages are now connected by *katcha* roads. One of the significant developments in the area is construction of two road bridges across the lake connecting Eluru and Kaikalur towns located almost at opposite ends of the lake. There is now a regular bus service operated by the road transport corporation. Post office, electricity and telephone have been provided to many villages during recent years. But even now, some of the villages are isolated in rainy reason. Boats fom the only means of transport for people during this season.

The landscape of Kolleru lake has been considerably changing during the last few decades. The lake which was at one time, smaller in size has become considerably large by becoming shallow due to silt deposition submerging vast areas in and around the lake. Fishing and hunting of birds were the earlier activities and there were only a few settlements with the lake area in the beginning. The starting of cultivation and consequent increasing human interference with natural environment, has disturbed the ecological balance.

Some of the present problems are the tidal effect in the lake perephery by the invasion of sea water through Upputeru during periods of high tides, silting up of the lake by sediments brought by streams draining into it, obstruction of free flow of water due to cross bunding by cultivators, earthern bunds of fishing tanks, extensive growth of weeds and some of the roads formed recently connecting one village with the other.

Another obstruction preventing the free circulation of water is the construction of bamboo fences (*dadikattoos*) across the lake for catching fish.

Pollution of water is another major problem. The effluents from sugar factories and sewage from the towns are the source of organic pollution lowering the dissolved oxygen content and affecting the survival of fish. Besides these, a lot of inorganic pollutants pertaining to fertilizers and pesticides carried into the lake through drains flowing from the surrounding agricultural fields are also hindering existence of fish. Drinking-water is also affected by these pollutants.

The natural environement created by the presence of flora and fauna is rapidly vanishing from the region due to interference of man through cultivation and fishing activities.

While preparing the perspective plan for solving all these problems and to achieve over all development of this lake region, planners should bear in mind the ecosystem of the region and see that the ecological balance is not disturbed.

PART—II
ALTERNATE ENERGY SOURCES

20

NON-CONVENTIONAL ENERGY RESOURCES

In view of rapid depletion of non-renewable resources like coal and petroleum, and unassured water supply to generate hydel power due to uncertain rainfall in several parts of our country, there has been much awareness and realisation to develop non conventional energy resources. The energy crisis has focussed on the urgent need for development of non-conventional energy needs.

Realising the gravity of the situation the fuel policy committee set up by Government of India has strongly recommended that all states should take non-conventional energy development programmes on a large-scale to meet the energy demand. Moreover the non-conventional energy resources being pollution free keep the environment clean. Being abundantly available, perennial and environmentally safe, the non-conventional energy resources have the potential to set forth a sustainable and equitable socio-economic development in the country. Efforts are being made in the last two decades under the policy and guidance of Commission of the Additional Sources of Energy (CASE) a Department of Non Conventional Energy Sources (DNES) was established in 1982 by the Government of India. This department has become a full fledged Ministry of Non-Conventional Energy Resources in 1992 (MNES). An impressive and rapid progress has been made during the 8th plan period (1992-93 to 1996-97). A variety of renewable energy systems have reached technological maturity and economic variability of wide range of applications. About 1050 MW of power generating capacity constituting 7 per cent of the total power capacity addition in the country during the eighth plan period has been solely attained by the non-conventional energy sources, mainly wind energy. Besides millions of rural people have been immensely benefitted from biogas plants, improved *chulhas* and solar lanterns.

India is implementing one of the world's largest programmes on renewable energy, which includes improved *chulhas*, biogas plants, short duration fuel wood tree species, biomass gasfiers, solar thermal and solar photovoltaic systems, wind farms, wind mills etc.

The MNES is allocating sufficient funds for research and development, demonstration, and rural oriented extension programmes. It is also extending institutional finance through its financing arm namely Indian Renewable Energy Development Agency (IREDA) for commercially viable projects with private sector participation. It is providing subsidies for individuals and institutions for installing devices and also arranges bank loans.

Some of the non-conventional energy resource development proggrammes like biogas plant are being implemented by organisations like Khadi and Village Industries Commission (KVS), Mumbai, National Dairy Development Board (NDDB) Anand Gujarat, All India Women's Conference (AIWC) New Delhi, South Asia Partnership India (SAP) New Delhi and Sustainable Development Agency (SDA), Kanjirapally, Kerala.

An estimate of the potential of important renewable energy sources are shown in Table 1.

TABLE 1
ESTIMATED POTENTIAL OF IMPORTANT RENEWABLE ENERGY SOURCES IN INDIA

Sources/Systems	*Potential*
Biogas Plants (Nos)	12 million
Improved *Chulhas* (Nos)	120 million
Biomass	17,000 M.W.
Solar energy	20 MW/sq. km.
Wind energy	20,000 M.W.
Small hydel power	10,000 M.W.
Ocean Thermal energy	50,000 M.W.
Urban and Industrial Wastes	1,700 M.W.

Biomass as a Source of Energy

The term biomass includes all kinds of plant species (fuel wood plants, trees, grasses, bushes, aquatic weed) of land and waterborn, animal wastes particularly cattle and buffaloes dung, agricultural crop residnes, industrial wastes like bagasse, husk, saw-dust, night soil, garbage, sewage solids, oil yielding petro crops, etc. All these varieties of biomass can be broadly grouped into forest biomass, agricultural biomass aquatic inland/marine biomass and human habitat biomass. The carbon content in biomasses is causing less air pollution while burning when compared to fossil fuels. The use of biomass as a source of energy has tremendous potential in our country. Out of total estimated quantity of 1500 million tons of biomass, 770 mil tons is available for energy conversion. The energy produced out of this quantity is equivalent to 77 per cent of present energy consumption in our country. At present 40 per cent of total energy consumed in India comes from sources like wood, wood wastes and animal wastes. Most dry farms of biomass can be burnt directly to produce heat, steam or electricity.

Animal Waste (Cattle and Buffaloes Dung)

Among the sources of biomass animal waste i.e. cattle and buffaloes dung is the most important bio energy resource. With vast potentiality, this is being utilized largely for biogas generation in our country which has largest number of cattle and buffaloes since it is basically an agricultural country. The estimated dung potential available is sufficient to instal 12 million biogas plants. Realising the availability of vast quantities of dung which has been wasted as an inefficient fuel in the form of dung cakes since several years, the Ministry of Non-conventional Energy Sources has started a massive programme of biogas development from cattle and buffaloes dung under the title of National Project of Biogas Development in 1981-82 to promote the establishment of biogas plants all over India.

Biogas is a cheap and clean fuel in the form of gas. It contains 55 to 75 per cent of methane, a gas which is inflammable. Biogas is generated through an organic process in a biogas plant and is also called as gobar gas. It is produced due to anaerobic digestion of organic matter like dung. The digested slurry is an excellent organic manure whose value is not diminished in the process, rather it has enhanced. Both fuel and manure are obtained from the same quantity of dung. Hence the biogas technology helps in generating economic benefits in terms of savings of firewood as well as chemical fertilizers, besides social benefits like removing drudgery of women and children in collecting and head loading of heavy bundles of firewood every day and making arrangements for their storage in the rainy season. It also eradicates certain ailments relating to eye and lungs due to smoke. It saves time in tedious cooking and cleaning of black utensils and improve sanitation and environment by checking deforestation for fuel wood needs. The heat generated by burning biogas is 20 per cent more than that of dung cakes. The slurry is 43 per cent more fertile than compost. It saves consumption of kerosene, reduces the quantity of firewood consumption, which is inadequate in our country. Moreover it is safe to use since the plant cannot break or explode unlike the cylinder of cooking gas, when handled carelessly.

About 27 lakh biogas plants of family size for domestic use for cooking and heating in individual households were installed so far, thereby covering 22.5 per cent of the estimated potential. It is estimated that the gas generated by these plants is equivalent to a saving of 50 lakh tons of fuel wood worth 200 crores of rupees per year. Besides, these plants produced 240 lakh tons of enriched manure, converting 2.4 lakh tons of nitrogen, potash valued 160 crores every year. During the year 1997-98 over 16.5 lakh plants were installed out of a target 17.5 lakhs.

Taking advantage of life span of the plants to be 20 years, large economic benefits would thus be generated to the nation. In addition, the programme has created employment of about 70-75 lakh man days per year of both skilled and unskilled workers in rural areas. The NPBD provides subsidy to the beneficiaries, which ranges from 1700 to 6200 depending upon the size of the domestic use biogas plants.

The biogas plants for family use are manufactured in sizes from 1 to 10 for the requirement of the number of people. The size of the plant is determined by the gas production in cubic metres per day. Though there are ten sizes of plants, there is demand only for sizes 1 and 2. Size No.1 requires 25 kgs of wet dung for which 2 to 3 cattle and buffaloes are required and the gas is sufficient for cooking needs of 3 to 4 people. Whereas size No. 2 requires 50 kgs of wet dung for this 4 to 5 cattle and buffaloes are required. The gas produced in No. 2 size biogas plant is meant to cater to the cooking needs of 5 to 8 persons. Size No. 10 which is the largest of all requires 250 kgs of wet dung which 20 to 22 cattle buffaloes are needed. The gas produced is sufficient for the cooking needs of 26 to 32 people. One kg. of wet dung generates 0.037 cubic metres of gas and a biogas burner for one hour burning consumes 0.28 cubic metres of gas. Lighting of one biogas bulb consumes 0.14 gas per hour. Biogas which generates from human excreta per head is about 0.028 cubic metres.

The state-wise achievements and targets of biogas plants installed for domestic use during the year 1997-98 are shown in Table 2.

TABLE 2
STATE-WISE TARGET AND ACHIEVEMENT OF BIOGAS PLANTS – 1997-98

State/UT/Agency	*Number of biogas plants*	
	Target	*Achievement*
Andhra Pradesh	16,000	18,929
Arunachal Pradesh	100	65
Assam	1500	138
Bihar	1500	641
Gujarat	12,000	12006
Haryana	2000	1300
Himachal Pradesh	1200	1142
Jammu and Kashmir	200	Nil
Karnataka	19,500	12,465
Kerala	1500	765
Madhya Pradesh	18,000	15,260
Maharashtra	15,000	15,406
Manipur	300	271
Mizoram	200	147
Orissa	7500	7164
Punjab	6000	5524
Rajasthan	1500	1526

State/UT/Agency	Number of biogas plants	
	Target	Achievement
Sikkim	250	174
Tamilnadu	2000	1084
Tripura	100	72
Uttar Pradesh	2000	8463
Andaman and Nicobar islands	10	5
Chandigarh	10	-
Dadra and Nagar Haveli	10	-
Delhi	10	-
Pondicherry	10	-
KVIC Mumbai	40,000	46,285
NDDB Anand	100	55
AIWC New Delhi	1500	1215
SAP New Delhi	2500	2145
SDA Kanjirapally	2000	2000
Total	1,75,000	1,65,573

The programme is implemented by state nodal departments and agencies, Khadi and Village Industries Commission, Mumbai; National Dairy Development Board (NDDB), Anand, Gujarat; All India Women's Conference (AIWC), New Delhi; South Asia Partnership-India (SAP), New Delhi; and Sustainable Development Agency (SDA), Kanjirapally, Kerala. The state nodal departments and agencies and KVIC are involving a large number of rural based voluntary agencies and enterprenuers in the construction of biogas plants on turnkey basis with free maintenance warranty for a period of three years.

Subsidy is provided in fixed amounts irrespective of size of plants and category of beneficiaries. It varies from Rs. 2000 to 5000. Highest subsidy of Rs. 5000 is given to North eastern region states (Sikkim and Jammu and Kashmir). For some hilly districts in UP, Himachal Pradesh, West Bengal and Tamilnadu and for SC, ST people and for people below poverty line a subsidy of Rs. 2500 is given.

Community Institutional and Night Soil Based Biogas Plants

The MNES has also launched a community and institutional and night soil based biogas development programme which has been implemented by various states. It is an integrated, decentralised waste recycling and energy regeneration programme for running dual fuel engines, for drinking water supply and for generating electricity.

The Community Biogas Plants (CBP) are installed for a group of beneficiaries who do not possess enough cattle individually but as a group of individuals of some households who possess enough cattle to meet the input for establishing biogas plants ranging from 15 M^3 to 83 M^3 capacity. The input, output is shared by the group of beneficiaries themselves and the plants are managed by Panchayats. In our country 401 community biogas plants were installed by the end of March 1992.

!nstitutional Biogas Plants (IBP) of sizes 15 M^3 to 85 M^3 which are installed in institutions, schools, *goshalas*, hotels, dairy farms called institutional biogas plants. These plants were maintained by the concerned institute. In our country 359 institutional biogas plants were set up by the end of March 1992. Besides these two categories, night soil based biogas plants (NBP) were also installed in some places in our country.

Biogas development programme has set a physical target of 300 plants of all three categories combined for the year 1997-98. Out of 300 plants, 263 plants have been completed. Highest number of these plants 48 out of 50 were installed in UP and Maharashtra stands second in this respect with 30 plants installed out of target figure of 35. The programme is implemented by state governments, state nodal agencies and NGOs. With a view to overcome social inhibitions attached with the use of biogas generated with nightsoil based biogas plants, NGOs have been specifically involved to educate the beneficiaries.

Subsidy is provided for North eastern region states and Sikkim for small size plants linked with toilet complexes and plants set up by charitable institutions. Besides, community, plants set up by SC/ST and families below poverty line were also given higher subsidy.

Community Toilet Complexes

Non Conventional Energy Development Agency (NEDA), UP has been implementing the programme of night soil based biogas plants on a large-scale and a total of 325 such companies have been set up since 1998-99. Mostly in areas of many cities like Kanpur, Lucknow, Agra, Ghaziabad and Varanasi. These plants have improved the environment of the locality and at the same time have helped in generating biogas to operate pumps for drinking water supply and lighting of complexes.

Toilet Linked Biogas Plants in Schools

Toilet linked biogas plants are promoted to provide sanitation in schools particularly women teachers and girl students and also to educate students on the diversified application of biogas for cooking and lighting purposes. In addition to 207 plants installed in 1996-97, 376 more plants were installed during the year 1997-98. Highest number of these plants (122) were installed in West Bengal during this year. The installation of these plants was mostly done by KVIC which set up 222 out of 376 installations.

Depending upon the type, size of the plant and category of people (general or special) and the region the subsidy varies. For CBP plants the subsidy in Rupees ranges from 2,25,000 (for 85 M^3 size) to 44,000 (for 15 M^3 to 20 M^3 sizes) for general category and 2,70,000 (for 85 M^3 size) to 2,10,000 (for 45 M^3 size) to 24,000 (for 15 M^3 to 20 M^3 sizes) for N.E. region, and special institutions like *goshalas*, charitable organisations, cooperative societies or trusts. For other states and private and profit making institutions the subsidy ranges from 1,40,000 (for size 85 M^3) to 15,000 (15 M^3 to 20 M^3 sizes). In the case of NBP plants the subsidy for N.E. region ranges from 4,60,000 (25 M^3 to 35 M^3 sizes) to 2,50,000 (for 15 M^3 to 20 M^3 sizes). For other states the subsidy for this plant varies from 8,00000 (60 M^3 sizes) to 70,000 (for 15 M^3 to 20 M^3 sizes).

Improved Chulhas Programme for Fuelwood Biomass Utilization Efficiency

Fuelwood, crop residues, dung cakes are used as fuel and are burnt in stone and mud made stoves called *chulhas* by rural population mostly and also by urban slum dwellers. Among these three fuel resources, wood is the predominant fuel resource which is largely used for cooking and heating purposes. A characteristic feature of the fuelwood is that 70 per cent of it is collected from forests, village plantations and other nearby sites. It is collected freely and as such it is not recorded. Collection is done by women and children. These *chulhas* are inefficient in burning wood, emiting lot of smoke causing respiratory and eye diseases. It has been estimated that 1 per cent increase in the efficiency of these traditional *chulhas* would save 3 million tons of fuelwood per year.

The Ministry of Non-coventional Energy Sources launched a national programme of improved *chulhas* in 1983 not only to increase the efficiency of *chulhas* by 20-25 per cent by adopting new designs but also check health hazards. By increasing the combustion efficiency of *chulhas*, lot of fuel-wood would be saved.

The programme has been undertaken with a view to conserve fuelwood, and forests to remove or reduce smoke in the kitchen, to check deforestation which causes environmental degradation, to reduce the drudgery of women and children from cooking in smoky kitchens and the collection of more fuelwood, to reduce the cooking time and save fuel and to generate employment opportunities to the rural poor by creating trained workforce of women and traditional craft persons competent to construct, instal improved models of *chulhas*. These improved *chulhas* make better living conditions of rural population and urban slum dwellers. These are also cheapest and best. The programme is being implemented by KVIC, state nodal departments and agencies, NDDB and AIWC all over India.

Under this programme three types of *chulhas* i.e. fixed, family size and community models and portable models of *chulhas* depending upon the requirements of beneficiaries have been designed.

Fixed Chulhas of Family Size Models

These *chulhas* has three principal components (i) A closed hearth where combustion of the wood takes place, thus protecting the fire from the vagaries of wind (ii) A chimney for bringing in air for combustion and removing the smoke from the kitchen (iii) one or multipot design–capable of carrying more than one pan and utilizing certain amount of heat from hot gases that would otherwise have gone waste.

The advantages of fixed model *chulhas* of family size are:

(i) It can be used in most of Indian homes,

(ii) It can be constructed by local trained persons,

(iii) It can make kitchen smokeless,

(iv) Two or three food items can be cooked simultaneously,

(v) It can be designed to suit the local needs of the population

(vi) It is easy to rapair and maintain,

(vii) It can be constructed with locally available material.

The department has approved 20 such models for popularization under the National programme.

Community Size Models

Community size models are fixed type of *chulhas* for large user's requirement. Two such models capable of cooking for 200 and 50 persons respectively were developed and installed at various institutions for demonstration. Based on feedback eight community models which have efficiency over 35 per cent have been approved for large-scale propagation. Community size model *chulhas* have been installed largely in schools/institutions, hotels, hostels, charitable institutions, etc, to cater the cooking needs of a large number of people in different parts of our country. These *chulhas* are capable of saving appreciable amount of fuel wood and are in great demand.

The programme has also included the development of damper-less models of improved *chulhas*. These *chulhas* are convenient for operation and the unit cost of *chulhas* has come down. The efficiency of these *chulhas* comprises favourably with that of *chulhas* with dampers. In view of the operation convenience and low cost, the damperless *chulha* models have been propagated from the year 1988-89 onwards.

Portable Chulhas

The portable models are made of mild steel, galvanized iron or cast iron and are suitable for the beneficiaries of urban/slum population which do not have sufficient space for cooking, migratory and tribal population which keep on changing their place of residence, and the rural people who prefer to cook outside the kitchen due to climatic conditions. The portable *chulhas* are manufactured by small enterprenuers in different parts of the country. Maximum thermal efficiency limit of the portable models was raised to 25 per cent from 20 per cent. Subsidy is given to beneficiaries, which range from 50 to 75 rupees depending upon the type of beneficiary and the type of *chulha*. The cost of fixed *chulha* varies from Rs.75 to 80 whereas it varies from 136 to 200 rupees in the case of portable *chulha*. In our country over 12.5 million improved *chulhas* have been installed by the end of March 1992.

A physical target of promotion of 30 lakh improved *chulhas* was fixed for 1997-98. Over 25.90 lakh *chulhas* have been installed statewise, and agencywise achievement and targets are shown in Table 3.

TABLE 3
NATIONAL PROGRAMME ON IMPROVED *CHULHAS* ACHIEVEMENT AND TARGETS

State/UT/Agency	*Target*	*Achievement*
Andhra Pradesh	2,00,000	2,57.360
Arunachal Pradesh	15,000	925
Assam	50,000	490
Bihar	20,000	2,878
Gujarat	90,000	90,812
Goa	5,000	4,077
Haryana	50,000	7,416
Himachal Pradesh	30,000	19,211
Jammu and Kashmir	50,000	
Karnataka	88,000	20.295
Kerala	1,00,000	24,948
Madhya Pradesh	2,50,000	1,88,969
Maharashtra	2,00,000	1,77,011
Manipur	15,000	3,155
Meghalaya	10,000	
Mizoram	15,000	3,000
Nagaland	10,000	157
Orissa	1,40,000	1,31,228
Punjab	1,00,000	59,500
Rajasthan	1,00,000	99,886
Sikkim	5,000	5,114

State/UT/Agency	Target	Achievement
Tamilnadu	1,00,000	99,460
Tripura	15,000	11,549
Uttar Pradesh	2,30,000	1,73,703
West Bengal	1,90,000	1,92,034
Andaman & Nicobar	1,500	1,402
Chandigarh	2,000	-
Dadra & Nagar Haveli	1,200	1,199
Delhi	2,000	-
Pondicherry	3,000	4,517
Lakshwadeep	300	122
KVIC, Mumbai	8,70,000	9,84,506
NDDB, Anand	12,000	12,762
AIWC New Delhi	30,000	13,650
Total	30,00,000	25,91,336

It is evident from Table 3 that Andhra Pradesh's achievement is much higher than the target figure. West Bengal which has highest number of installed *chulhas* also crossed the target and stands second in achievement after Andhra Pradesh. The KVIC has installed highest number of *chulhas* and has more than crossed the target figure.

Apart from social and health benefits, improved *chulha* was found to be adequate to recover the entire cost of installation within two months of its use.

Biogas Fiers

Another device producing biogas by using biomass wastes is biogasfier. In a biogas fier device biomass wastes such as wood waste, wood chips, twigs, cotton stalk, *tuar* stalk, maize cobs, nice straw, saw dust, etc. would be converted into biogas. Gas is produced through chemical conversion processing. The gas runs the engine to produce motive power. In turn it runs pumpset for irrigation or drinking water or to produce electricity. The biogas fier pumpsets are available in 5 HP to 10 HP for mechanical applications like water pumping, from 3 KV to 100 KW for generating electricity. These systems substitute or replace diesel fuel in conventional diesel engine for respective applications by over 65 per cent.

The use of biogas fier as a supplementary fuel in existing diesel engines when operated on dual-fuel model appears attractive as long term strategy. These are well suited for water pumping for irrigation purpose and rural electricity generation. Biogas fier systems can vastly improve the quality of life in rural areas and provide employment.

The Ministry of Non-conventional energy sources has also taken up a programme of biogas fiers. In our country biogas fiers produced 7.9 MW of power till the end of December 1992. A total number of 52 biogas fier systems with an aggregate capacity of 5670 KW equivalent have been sanctioned during 1997-98. But 28 biogas fiers have been installed during the year 1997-98 producing 2495 KW of power. Highest number of biogas fiers (27) were sanctioned to Andhra Pradesh out of which 10 were installed and this is the highest when compared to other states. Table 4 shows state-wise list of biomass gas fier systems installed and sanctioned during 1997-98.

TABLE 4
STATE-WISE LIST OF BIOGASFIERS INSTALLED AND SANCTIONED DURING 1997-98

State	*Sanctioned*		*Installed*	
	No.	*Capacity in KW*	*No.*	*Capacity in KW*
Andhra Pradesh	27	3040	10	1000
Assam	1	300	-	-
Gujarat	2	10	1	5
Karnataka	10	1820	5	860
Madhya Pradesh	1	100	-	-
Tamilnadu	11	400	7	130
West Bengal	-	-	5	500
Total	52	5670	28	2495

The Non Conventional Energy Development Corporation of Andhra Pradesh (NEDCAP) has installed 10 biogas fier systems each of 100 KW capacity in rice mills. It is estimated that one rice mill would be saving about 9500 litres of diesel oil by using gasfier system for a period of six months in a year. Tamilnadu Energy Development Agency (TEDA), Chennai made special efforts in promoting gasfiers for cooking purpose in central prisons at Chennai, Coimbatore, Trichy, and Vellore. Each prison would be saving about 11.25 tonnes of fuelwood per month by using a gasfier system of 50,000 Kcal/hour capacity.

Solar Energy

The sun provides enormous amount of energy in the form of solar radiation. Out of total energy radiated, 47 per cent only reaches the earth. The amount of solar energy received by earth varies and depends upon the angle at which the sun rays fall on the surface of the earth, duration of the day, time of the day, distance of earth from the sun on the orbit of its revolution around the sun, changes in the output of solar energy and transparancy of the atmosphere. On an average 500 calories of energy in a square centimetre area per day is received in our country. In Delhi, it is about 364 calories in winter, 576 calories in summer per day.

In respect of annual distribution of sunshine in our country it varies from 2000 to 3200 hours depending upon the region.

Application of solar energy can be broadly divided into conversion of solar energy into heat, and conversion of solar energy into electricity. As such, for utilizing solar energy in our country two programmes, i.e., solar thermal energy programme solar photovoltaic programme have been taken up by the Ministry of Non-conventional Energy Sources under whose guidance all the states are implementing these programmes.

Solar Thermal Energy Programme

Under this programme, solar thermal technology is based on thermal conversion of solar energy using solar collectors. This programme is aimed at supplementing thermal energy requirement at various temperatures by harnessing perennial, non polluting and freely available solar energy by directly converting into energy using solar thermal devices and systems. solar thermal energy systems have been broadly categorised as:

(i) Low temperature systems like water heaters, solar driers, solar air heaters, solar timber kilns, solar desalination systems, (Stills), solar green house, etc., having maximum temperature upto 100°C.

(ii) Medium temperature systems like solar cookers,

refrigeration, airconditioning, solar pumping, industrial process heating, etc., having temperature of operation between 100°C and 250°C.

(iii) High temperature system suitable for solar thermal power generation where operating temperatures are above 250°C.

Solar devices and systems which have been listed above are in various stages of development and commercialisation. These devices are used in rural and urban areas both in domestic and industrial sectors.

Solar cookers

Under solar thermal energy devices, solar cooker is one of the most important device. It is a device meant for cooking. Several types of solar cookers have been developed. These are (i) simple box type cookers, (ii) multireflector type cookers, (iii) concentrating type cookers. Depending upon the type of cooker, temperatures in the range 120°C to 300°C can be attained. The hot box type or open type solar cookers can cook many type of dishes which need low and slow heating such as rice, vegetables, baked items, cooked dishes, etc. They cannot cook items requiring high temperature such as *chapatis*, or fried dishes. Even in respect of some of these items they can be used to make the food articles tender which can be finally cooked on conventional fuel stoves in lesser time so that the amount of conventional fuel can be saved. Thus, cookers can save 30 to 50 per cent of commonly used cooking fuels like wood, coal, LPG, Kerosene, etc. If used properly and regularly concentrating type of cookers can attain high temperatures depending upon concentration ratio/factor. The drawback with such cookers is that cooker has to be directed towards the sun after 10-15 minutes and if automatic devices for such tracking are provided, the cost increases. All types of solar cookers including concentration types can be used depending upon the application. In 1982, India became the first country in the world to start regular large-scale commercial manufacturing and marketing of solar cookers. In the 1950s a solar cooker was developed in India but not marketed on a large-scale due to its high cost.

Drawbacks. There are few limitations which prevent the popularization and large-scale use of solar cookers. One is it has to be kept in an open place where sunshine should be available. Since the houses are too close and without much open space in villages and urban slum areas there is absence of sunshine due to shade. Secondly, it cannot be used during nights and cloudy and rainy days. Lastly working women who have to leave their homes in the morning and are away for the major part of the day, are not able to use them.

In our country 2,38,906 solar cookers were distributed by the end of December 1992. Solar cooker which requires normally 2 hours for cooking for a household, is equivalent to burning of 1 kg. of wood in a stove for cooking.

Solar Water Systems

Another important solar thermal energy device is solar water heating system which is meant for heating water. The system consists of a flat plate solar collector and a storage tank. The collector is connected to a double walled insulated storage tank containing water. When water is filled in the system, and the blackened absorbor sheet of the collector is exposed to the sun, it gets heated and the temperature of water increases, since it is in contact with the hot sheet.

As the temperature of water increases it tries to remain in the top, hot water gets collected in the storage tank cold water being drawn in from the bottom. Solar water heating systems have many applications in the domestic and industrial sector. For providing solar water heating system, the initial investment is more. As such the users are not readily coming forward to take up these systems even though sufficient amount of subsidy is provided. It may be made compulsory for multistoreyed buildings, hotels, etc. There are a number of industries which use hot water. While giving licences it may be made compulsory to instal solar heating systems for those industries which require hot water for processing. In our country about 10,127 domestic solar heating systems were installed by the end of March 1992.

Solar Stills (Solar Desalination Plants)

A solar still also called as solar desalination system works on the water heating principle with the difference that in this case water is kept in a blackened metal/plastic tray covered with sloping transparent roof which is exposed to solar radiation. In other words water does not flow through tubes/channels but it is stagnant. When it gets heated up some water gets evaporated (leaving the salt behind) and vapours strike the sloping roof and condense into water drops. The pure water drops thus formed are collected through channels. About 3 to 4 litres of water can be obtained from one square metre area of the system per day. The solar desalination system can be used to provide water for drinking in areas where only salty or brackish water is available. It can be also used to provide distilled water needed for batteries. In our country 9651 solar stills were installed by the end of March 1992.

Solar Driers and Timber Kilns

The solar driers which are also called solar air heaters can be used for various applications like drying food grains, vegetables, wood, etc. Products dried in solar driers are as good in quality and food value as compared to those in conventional driers. In solar air driers air is blown through the tubes/ channels or to go over or below the blackened absorber sheet. Temperatures as high as 130°C can be easily attained with simple systems. The hot air can be utilized to dry any material such as wood, or agricultural crops with increasing speed and efficiency of such drying several times more than the traditional method of direct exposure to sun. The heated air can also be used to operate an engine. In our country 61 solar driers were installed by the end of March 1992.

Solar timber kiln is another solar thermal device. Only seven devices were installed in our country till the end of March 1992.

Solar Photovoltaic Systems Programme

Direct conversion of solar energy into electricity can be achieved by using photovoltaic effect in certain materials. The basic unit of photovoltaic system is solar cell. While the actual system prices are still not competitive compared to those based on conventional energy sources, there are hopes of further sizeable reduction in costs in view of increasing role of photovoltaic systems in the energy scene in future especially in developing countries.

Solar photovoltaic systems have several advantages such as modulatory, ease of installation, low cost maintenance free from noise pollution, free operation, etc. They can be utilized to meet the electrical energy requirement for a variety of applications. The technology is considered to have significant potential in India and is particularly suitable in remote and isolated areas which are not served by conventional electricity grids. The Ministry of Non-conventional Energy Sources has taken up the development, production and application of solar photovoltaic devices in our country. Street, domestic, community lighting, colour television, pumpsets, lanterns are the various photovoltaic systems which have been distributed in several states in our country.

In our country 8100 villages have been provided with photovoltaic street lighting systems and 9050 domestic lighting systems, 968 community and television systems by the end of March 1992.

Solar Water Pumpsets

Solar water pumping systems are also having significant potential in India especially in electrified villages. Such pumps have been found to be technically feasible for shallow and deep well applications. These pumpsets can be installed in the irrigation wells for drawing water with the help of solar energy. Each pumpset of 1 KW with cost about 2.5 lakh rupees.

The main constraint for wider use of solar pumps have been the high initial cost of such systems. Considering the likelihood of reduction in the cost of photovoltaic modules and the long term potential of solar pumping systems, a five year programme for gradual introduction of such systems for irrigation was drawn up to pave the way for eventual commercialisation.

This is to be achieved through a process of technology improvement, volume production, incentives, finance arrangements and expanding market. Different sizes of pumpsets are proposed to be offered to the farmers. The pumps are made to be available at a reduced price along with financial arrangements. In our country there were 1171 photovoltaic pumps by the end of March 1992.

Solar photovoltaic systems are now commercialised by organisations like railways, department of telecommunications, and oil and natural gas commission. The department of telecommunications is progressing ahead with its programme for rural radio communication systems powered by solar photovoltaic modules. This has led to a large commercial market for Indian photovoltaic industry and also established the technical and economic viability of photovoltaic systems for low load and high reliability applications.

Wind Energy

Wind is also a source of energy which varies with the speed. Higher the speed of the wind, greater is the energy generated. Higher amount of energy that can be harnessed from wind also depends upon volume. Wind energy is renewable and poses no environmental threats. Wind speed and their directions vary constantly throughout the day and night and from one geographical location to another. These variations may be smaller or larger depending upon the season and geographical location. As such there will be fluctuations in wind energy output also.

Wind energy has been used for thousands of years to propel ships and for pumping water and grinding grain. Tapping of wind energy dates back to nearly 200 BC when the Persians lifted water from wells for drinking purpose. Wind mills were used in Netherland for lifting water from low lying areas. There were about 10,000 wind mills in Netherlands in the 19th century. With improvements in design wind mills were first used for electric generation in 1980 and in a commercial scale in 1940 in USA. Where the wind speed is not high the wind mills should be set up at greater heights and they should have larger wings.

In our country preliminary estimates have revealed that wind power potential would be over 20,000 MW. Indian winds are seasonal. Wind speeds all over India are higher during monsoon months. Even then there are certain areas in our country which are ideal for setting up wind mills.

The estimated wind energy would come to 1/3rd of the total hydel and thermal energy generated in our country. Half of this can be generated in Tamilnadu and Gujarat which are the most ideal states for wind power generation due to high speed winds.

To harness such a large source of energy, a broad based wind energy programme has been designed by the Ministry of Non-conventional energy resources. The programme basically comprises the assessment of wind, potential areas by conducting intensive surveys, development of efficient technologies for harnessing vast wind potential in the country for various applications such as wind power generation, water pumping and battery charging.

Under wind potential assessment programme, wind monitoring and mapping activity has been taken up both in the plains and on hills and mountains. The information is most useful in planning large grid connected electric generators known as wind farms, stand alone wind electric generators, battery chargers and wind pumps in wind potential areas.

Indigenous production and commercialisation of wind energy systems is a major thrust area of wind energy programme during the eighth plan. Significant progress was achieved under this programme. In our country 2983 wind pumps, wind farms generating 43025 MW of power and 107 wind energy battery chargers have been located till the end of December 1992.

Wind Farms for Power Generation

A wind farm is made of several wind energy generators arranged in a cluster whose combined output is fed in a grid. The most important are those located at Okha, Mandvi in Gujarat, Puri in Orissa, Deogarh in Maharashtra, and Tuticorn in Tamilnadu.

For power generation, areas that have wind speeds of 18 KMPH and above are chosen. In a plot of 1 sq. km. 90 wind mills with a total capacity of 5 MW of power can be set up. They produce 5 million units of power. Wind energy is a mechanical energy which can be converted into electrical energy with minimum losses.

The space required for erecting a single wind mill will be only 100 sq. metres (10m x 10m). However, if a group of wind mills is to be installed at the same location then the distance between two wind mills in a row will have to be 5 times the diameter of the rotor and the distance between the successive rows has to be 7 times the diameter of the rotor. The approximate land required for setting up of 1 MW capacity of wind farm is about 20 acres. This will marginally differ depending upon the topographical conditions. The present day capital cost towards the installation of wind farms is estimated at about 3.5 crores per MW. This figure is an estimate, but the actual cost would depend upon the size, and number of machines to be installed in a site and length of distribution line to be laid to interconnect farm with the grid.

Wind energy electric generators need no fuel and therefore there will not be any escalation in the cost of energy. It will be cheaper if the tax benefits are taken into account. It is generally accepted that the cost of generation from wind is cheaper than the cost generated by diesel. In the long run the cost of generation may even be cheaper compared to the cost of generation from conventional thermal stations since it does not suffer from any escalation. In our country 43.05 MW of power generated wind farms were put into operation by the end of March 1992.

Wind Energy Water Pumps

Apart from power generating programme, wind energy water pumping programme has also been taken up by all the states for irrigation and drinking water purposes. The wind mills attached to a pump can lift water from wells and storage tanks. Wind mills are extensive in many parts of the world. They hold a significant potential for pumping water for irrigation, drinking water needs, etc., in rural areas having low or moderate wind regimes, especially in developing countries.

Under demonstration programme, an aggregate capacity of 51 MW has been established at 23 locations in 8 states by the 31st March 1998. A total of 390 million units have been fed to the grid from these demonstration projects. In the demonstration projects undertaken so far the unit size of machines has gone up from 55 to 100 KW in the first few projects to 400 to 600 KW in recent projects.

The commercial projects are being set up with the assistance from financial institutions and through private investments. Indian Renewable Energy Development Agency (IREDA), the corporate financing arm of Ministry of Non-conventional Energy Sources provides loans for wind power projects. International financing is also available to IREDA under World Bank. Direct international financing is being arranged for commercial projects. The total power generated under commercial projects programme was 66.83 MW by the 31st March, 1998.

State-wise wind power installed capacity (in MW) as on 31.3.98 is shown in Table 5.

TABLE 5
STATE-WISE WIND POWER INSTALLED CAPACITY (IN MW) TILL 31.3.1998

State	*Demonstration Projects MW*	*Commercial Projects MW*	*Total Capacity MW*
Tamilnadu	19.355	687.940	707.295
Gujarat	17.345	149.565	166.910
Andhra Pradesh	3.050	52.740	55.790
Karnataka	2.575	14.435	17.010
Madhya Pradesh	0.590	11.700	12.290
Maharashtra	4.600	0.995	5.585
Kerala	2.025	-	2.025
Orissa	1.100	-	1.100
Others	0.465	-	0.465
Total	51.105	917.375	968.480

It is evident from Table 5 that Tamilnadu state stands first in the installed capacity both in the case of demonstration and commercial projects and Gujarat stands next after Tamilnadu in the production of power. Wind power generation is confined to a few states since wind velocities are not favourable in several states.

In sum, non-conventional energy may have a bright future. In fact, this energy may become an inevitable necessity in view of the rapid depletion of energy sources like coal, diesel, petrol, etc. Even the hydel power is erratic in generation due to vagaries in monsoon rainfall. Moreover our population is growing enormously. Keeping all these facts in view, the government of India is taking measures to develop non-conventional energy resources on a war footing. To meet several constraints in the utilization of these resources, the research and development wind of MNES has been carrying out considerable studies to improve the systems. If the cost of unit of power generated under solar photovoltaic system comes down, the energy demand in our country could be met to a large extent. Development of decentralised and integrated energy system both in urban and rural areas will ultimately solve the energy crisis in our country.

21

GEOTHERMAL ENERGY: A SUPPLEMENT TO CONVENTIONAL ENERGY

Geothermal energy is the heat energy stored in the sub-surface or in the interior of the earth. Internal heat is the dominant component of earth's energy system and it acts as a primary drive force for conversion in the earth's interior. It controls also the physical state and tectonic processes of the lithosphere. As such the surface heat flow pattern is controlled by the structure and dominant processes within the earth. Knowledge of regional heat flow pattern is therefore an essential input for understanding the evolution of the tectonics of the earth's lithosphere system. The first surface heat flow measurements were made by E.C. Ballard in 1939. The surface heat flow data combined with geological, geophysical and geothermal knowledge not only reveals the tectonic nature of a segment on the earth but also enlightens dynamic processes inside the earth that affect a region. Manifestations of earth's heat at many areas are commonly known as geothermal fields which provide opportunity for exploring geothermal energy to supplement the conventional energy.

The temperature increases from the surface of earth towards its interior very rapidly and at a depth of 50 km., it is 1200°C which is beyond the melting point of rocks. As such enormous quantity of heat is stored in the earth's interior. The enormous quantity of stored heat energy has caught the attention of Geoscientists who conducted lot of scientific studies for exploiting this energy partly atleast from a few thousand feet depth for utilizing it for certain industrial and domestic uses.

Before knowing about the scientific studies undertaken by geoscientists for tapping geothermal energy it is necessary to understand how much of heat is stored in earth's interior. Most of us are aware that earth being one of the planets of solar system was originally in a gaseous mass about 6000 million years back. It slowly cooled and reached the liquid state in 1500 million years, i.e., 4500 million years back. Since it was continuously cooling the outer layer with a thickness of 30 to 60 km. solidified due to radiation of heat. The outer layer is called the outer crust or lithosphere. The surface of the outercrust on which we are living had rapidly cooled due to radiation of heat into atmosphere. The outermost layer is also called sial since it is mostly composed of silica and aluminium. The next layer below this layer is called sima consisting mostly silica and magnesium and it is denser than the outer layer and the temperatures are also very much higher in this layer. All the continents belong to the outer crust and they are supposed to be floating on denser sima which is supposed to be in a liquid state but it is kept in a semi-solid state since the melting of rocks is raised under very high pressure existing there under compression. Though the heat had radiated from the surface of the outercrust and cooled but much of the heat from subsurface or earth's interior could not escape as such very high temperatures are maintained there.

The heat from earth's interior reaches the surface of outer crust broadly due to tectonic and volcanic activities. It is manifested through fumeroles which emit steam, hot springs, geysers which emit hot water and steam alternatively at regular intervals, salfataras emitting sulphur gases, and huge quantities of lava. All these features are characteristic in volcanic regions. The lava which is nothing but liquid rock maintaining high temperatures, comes out from sima layer where it is known as magma and the characteristic rock is basalt. The other endogenetic or internally originated activity is diastrophic activity which is associated with vertical or horizontal movement of earth's crust resulting in mountain building or continental drifting or sliding. The mountain building phenomenon is also called tectonic or orogamic activity. All the crustal movement activities confine to outer crust only. These movements result in faulting i.e. displacing of rock strata along a line of fracture, folding due to lateral thrust, fracturing, etc. When such crustal movements take place vibrations start, leading to earthquakes of various intensities. When such crustal movements take place, the equilibrium maintained between pressure and temperature is disturbed and the magma which is kept in a potentially solid state becomes liquid and expanded and reaches the surface through weak spots like joints and fracture zones under high pressure along with vast quantities of stream which is released from magma and this phenomenon occurs in volcanic regions. In tectonically activated regions, thrust faulting, and horizontal movement of continental blocks cause much friction and generate lot of heat. The water in contact with heat reaches surface in the form of hot springs or geysers. Part of the water reaches earth's interior from surface through percolation through joints, fissures and partly derived from magma.

Geothermal heat has so far been exploited by hydrothermal systems in few places of the world. Worldwide, the installed electric power from hydrothermal systems is approximately 2300 MW. Since 1970, proposals were made for tapping the so-called hot rock geothermal energy which is stored in hot crystalline rocks. This is being investigated in numerous research projects in USA, Japan, UK, Sweden, France and Germany (FRG).

The heat contained in the earth can be estimated at 10^{17} mega watt years (MWa) an incredible reservoir of energy. Yet the technical utilization of energy feasible from earth's crust is faced with tremendous problems and limitations. The present drilling technology permits drilling to a depth of 10 km. at an enormous expense. If a rock mass of 1 km^3. of earth's crust is cooled by 100°C thermal energy of 6800 MW would be released. This energy is more or less equivalent to the heat capacity of 6 million tons of oil. At sufficiently high rock temperature over 200°C this would be enough to operate an electric power plant with a capacity of 50 MW for 30 years. The electricity produced now from hydrothermal systems is only with temperatures over 150°C. Out of 2300 MW of electric power generated from hydrothermal systems 1000 MW are generated from California, 420 MW from Italy and 192 MW from New Zealand owing to exceptional situation as part of volcanic belt around the Pacific. Japan has been using geothermal heat for baths from times immemorial. At present 6000 MW of thermal capacity is being extracted world wide from hydrothermal sources with temperatures less than 150°C for heating and bathing purposes.

Geothermal Energy Prospects in India

Keeping in view the severe energy crisis which is going to occur in the near future as a result of rapidly growing population in India, much attention has been paid by the planners and scientists to exploit alternate energy resources to supplement the existing conventional energy potential. Geothermal energy is one such resource which has received much attention for exploitation since 1973 and sufficient scientific studies relating to places of occurrence, geotectonic associations, geochemical and thermal characteristics were carried out by scientists. On the basis of these studies, certain geothermal areas/fields have been identified, giving energy potential of each area and prospective uses.

Even though geothermal energy is manifested to the surface mostly in the form of hot springs, in several areas the degree of heat potential, chemical composition of water, and quantity vary greatly. Geothermal reservoirs emitting hot water with high temperatures normally occured in volcanic regions during the Pliocene and Holocene periods of quarternary era. Such regions occur mostly along continental plate margins and in some inter plate regions.

The geological history of India reveals that except the large-scale volcanic activity in the Deccan plateau during the tertiary period, no such activity during quarternary/recent periods is known to have occured anywhere except in Barren and Narcondom islands of Andaman and Nicober group. However, the movement of peninsular plateau towards north and its collision with Eurasian land mass with consequent under thrusting has played a major role in providing favourable environment for the development of geothermal resources of various grades.

The geological history of the earth reveals that all the present continents belong to the lithospheric sial block and were united together till the end of the Permian period. The composite block was called by Alfred Wagener, the author of continental drift theory as Panguea. During the triassic period the composite block was split into two major blocks, the northern block comprising Eurasia and North America was called as Angara land, and the southern block comprising, Australia, Antarctica, South America, South Africa, Peninsular India was called as Gondwana land by Wagener. The trough in between two blocks is called geosyncline and occupied by sea which was called Tethys. All these major continental blocks were subsequently called plates which are supposed to be floating on a viscous substratum of basaltic magma. The Gondwana land was further disrupted into independent continents in later periods. Peninsular India separated from Australia moved towards the Eurasian land mass. These continental plates are moving over basaltic substratum due to convection currents generated there. Peninsular Indian land mass moved towards north away from Australia after getting separated during 135 MYBP. Having moved towards north it collided with Asian block initially in 55 MYBP and finally in 40 MYBP.

Changes in Indian Subcontinent During the Tertiary Period

Important changes took place in the Indian subcontinent during the tertiary period. These are the landmarks in the geological history of India. One of the changes was that the north-western part of Peninsular India, which is called the Deccan plateau was subjected to intense volcanic activity and large quantities of lava reached the surface during 65 MYBP. The second remarkable change was the compression, thrusting of an elongated ocean trough called Tethys Geosyncline existing south of Asian land mass and compression of sediments in Geosyncline and raising of a folded mountain system i.e. Himalayas in the northern border in upper tertiary period. The third change was down faulting i.e. (displacement of rock strata along a line of fracture) of a narrow strip of land mass along its western edge in the late Pliocene period. Lastly block movements and reactivation of some ancient faults took place. The process of collision and compression gave birth to lofty mountain system which sprang up along the advancing edge of peninsular land mass in the north and to the compression of sediments deposited in the sea of Tethys. The mountain building activity along the northern margin of Indian peninsula had led to the formation of a belt of persistent subsidence between the central part of the shield and the rising mountain. The Sivalik mountains were the uplifted sediments deposited in the depression by Indogangetic and Brahmaputra rivers after draining the lofty northern Himalayas. It is now well established that the northern edge of Peninsular India was also involved in mountain building. As such, most of the lesser Himalayas belong to Peninsular India and the high northern chain of Himalayas belong to Tethys Geosyncline. The origin of Himalayas is linked to continent-continent collison, which created compressional forces acting towards north and which also caused compression buckling, squeeze and upward movements of its rocks. The recumbent foldings, with the axis towards south and the nappe structures being pushed on to the Indian land mass are the evidences for the compression.

As a result of prolonged northern movement and collision, large compressional forces at the corners of the subcontinent produced the syntaxial bends and one such bend is noticed in Kashmir. Anatexis also produced in certain parts of Himalayas producing mountains such as Nanga Parbat. The Mikar hills, Shillong plateau and the Ranchi plateau were uplifted and tectonically disturbed some other parts also. Partial melting took place in the crust under Tethys and higher Himalayas.

The collision of Indian Peninsular land mass with European land mass reactivated certain ancient faults and produced block lifts and epereogenic movements. In the beginning of cretaceous period, the Cambay area in western India experienced tensional forces oriented in an east-north east, west-south west directions which gave rise to Cambay lift. Ancient deep fractures such as the Narmada-Tapti-Son lineament continued to be rejuvenated. These are probably regions in which concentration of heat energy has occured and will continue to occur at shallow crustal levels. Seismic activity continues to occur and mountain building or organic activity is continuing in the Himalayas. The crustal temperatures under Higher Himalayas and Tethys Zones are comparatively higher when compared to peninsular land mass. There are possibilities of occurrence of zones of melting at shallow depths due to release of pressure on account of uplift and by seismic stresses under sialic layers (the outer crust of the earth or lithosphere). Such areas may be potential areas of geothermal energy in both peninsular India and Himalayan region.

Studies conducted by the scientists of National Geophysical Research Institute (NGRI) in these areas revealed that mostly non-volcanic geothermal resources of moderate to low grade can be expected in India. However, there is also a possibility of occurrence of high grade geothermal energy in some regions.

Over 300 natural thermal springs with temperatures varying from few degrees to over boiling point of water corresponding to the altitude of the place of manifestation, have been found in several parts of India. These occur in clusters or groups which follow certain tectonic trends and are closely related to regional and local fault systems thus making the development of hydrothermal systems possible.

Major Geothermal Regions of India

Based on tectonic and thermal histories, the following seven major geothermal regions have been identified after conducting studies by the scientists of NGRI with regard to occurrence of geothermal water manifestations. In each region the use of this energy for various purposes have been also mentioned.

The Himalayan Geothermal Region

Thermal springs in Himalayan region manifest themselves mostly along the banks of certain rivers or their tributaries. They are found even in Karakoram-Kailash mountain region beyond the northern Himalayas. In fact a belt of thermal springs starting from north-western Himalayas continues through Nepal and Bhutan towards north-eastern Himalayas to Burma and finally to Barren island. Due to continuing uplift of Himalayas, a large fracture system has developed in them, which facilitated the emergence of thermal springs, of which some have even high temperatures. Some of them have mild geyser activity emitting steam at regular intervals alongwith hot water. The temperatures range from 120°C to 240°C $\pm$ 20°C.

In this Himalayan geothermal region, seven sub-regions, i.e., Puga valley, Chumtong in Ladakh, Parbati, Beas and Sutlej valleys in Himachal Pradesh, Topoban region, Ala Kanda valley in Chamoli district of UP, and Kashmir valley basin have been also identified on the basis of geothermal characteristics of each sub-region and the use of energy for various purposes in each sub-region.

In Puga valley of Ladakh with estimated reservoir temperature of 240°C $\pm$ 20°C and with observed temperature of 84°C of thermal springs it has been suggested that the heat energy can be utilized for space heating, extraction and refining of borax, refining sulphur, generating electricity of 1000 kW using binary cycle, poultry farming, cesium extraction, mushroom cultivation and sheep farming.

In Chumtong region of Ladakh where geothermal reservoir temperature is about 180°C $\pm$ 20°C and the observed temperature of thermal springs is 87°C the hydrothermal energy can be utilized for green house cultivation, electricity generation, space heating, poultry farming, mushroom cultivation and fish culture, etc.

In Parbati valley, Kulu of Himachal Pradesh where the estimated reservoir temperature is about 160°C $\pm$ 20°C and the observed temperature of springs is about 96°C, the energy can be utilized for generating electricity, space heating, food processing, poultry farming, green house cultivation, etc. In Beas and Sutlej valleys of Kulu the estimated reservoir temperatures are about 120°C $\pm$ 20°C and the observed spring temperatures are 57°C and 60°C respectively. In both the regions the energy can be used for space heating, cold storage, poultry farming, pasturization of fruit juice and health resorts.

In Tapoban region of Alakananda valley of Chamoli district of UP the estimated thermal reservoir temperature is about 160°C $\pm$ 20°C and the observed spring temperature is about 64°C. The energy can be used is this region for electricity generation, baths, space heating, poultry farming, tourism, etc.

In Kashmir valley where the estimated reservoir temperature is about 100°C $\pm$ 20°C and thermal springs temperature is about 56°C the thermal energy can be utilized for space heating for small scale industries like wood processing, carpet, shawl, sari and other garment making, wooden craft, tourism, etc.

The West-Coast Konkan Region

A chain of mountain running parallel, to the west coast of India between latitude 16°–30° N and 20° N have thermal springs mostly in groups at 21 different locations. These are controlled by shear zones, fractures or dykes of Deccan trap and manifest in various river basins along stream courses. Many earthquakes occur in the west coast. The hot springs of this belt all are oriented in N-S and NNW-SSE directions indicating that the belt has suffered strong crustal movements and the fractures which control the thermal springs are still active. The temperatures of geothermal systems of this belt are around 100°C ± 20°C.

There are two sub-regions i.e. Unhavāre (Khed) and Ganeshpuri-Akloli in this region. In both the regions the estimated reservoir temperature is about 110°C ± 20°C and the observed temperatures of springs are about 70°C and 57°C respectively. The energy in both the regions can be utilized for refrigeration mushroom cultivation, drying fish, vegetables, and brewing.

Narmada-Tapti Geothermal Region

Thermal springs mostly in clusters with temperatures varying from 32°C to 98°C are found at sixteen locations in this region which is a permanent ancient geological feature of the peninsular India between 21° N and 24° N in a NNE-SSW direction. The estimated reservoir temperature of this region range from 60°C to 110°C except in one case, that is Tatapani Geothermal field (where hot springs having temperature is upto 98°C) whose estimated reservoir temperature is 160°C ± 20°C.

In Tapani, Surguja district of M.P., where the observed temperature of hot spring is about 98°C, the thermal energy can be utilized for electricity (installation of 20 KW binary cycle power plant is planned), sericulture, tourism, drying and curing of light, aggregate cement slabs, purification of bauxite, etc.

In Tapti basin of M.P. where the estimated reservoir temperature is 110°C ± 20°C and the thermal spring temperature is about 60°C, the energy can be utilized for electricity generation, sericulture, refrigeration, and processing industries.

The Damodar Graben Region

There are a few hot spring areas in Gondwana grabens adjacent to Damodar river ralley all along and the estimated reservoir temperature is about 100°C ± 20°C and the observed temperature of springs is upto 80°C. The energy can be used for electricity generation, rice para boiling, drying of agricultural products and paper pulp manufacture.

Combay Graben Geothermal Region

In the Combay and Kathana fields, the estimated reservoir hot eruptive springs (mixture of water and steam) of temperatures over 100°C occur. The energy from these springs can be used for electricity generation, refrigeration, processing and drying industries.

Delhi-Mobile Belt Geothermal Region

Springs in Sohna valley in Haryana with estimated reservoir temperature about 90°C ± 20°C and observed temperature of 47°C are found. These springs are found along the pre-Himalayan lineament towards south-west of Delhi with the areas of Delhi folding. The energy can be used for refrigeration and tourism.

Wardha-Pranahita Godavari Valley Graben Geothermal Region

The valley of the river Godavari along with those of Pranahita and Wardha rivers represent a prominent linear graben structure of Indian land mass. It is one of its main Gondwana basins of Permo-carboniferous to lower cretaceous geological periods. Most of its hot springs have temperature slightly over mean annual air temperature. Its hottest spring Agnigundala occurs in the bed of river Godavari and its temperature reached upto 62°C. Hot water also has been tapped through holes drilled for exploration of coal and ground water resources in various parts of this graben.

In Godavari and Chintalapudi sub-basins, the estimated reservoir temperature varies from 100°C to 156°C and the observed temperature of springs from 29°C to 62°C. The energy can be utilized for rice para boiling, drying of paddy, chillies, purification of bauxite mineral and cold storage.

Out of all these, four have been identified by NGRI scientists as good sources of geothermal energy.

Classification of Geothermal Systems

On the basis of estimated reservoir temperature and pressure, geothermal systems have been classified into five categories of HGS (Hydro thermal systems) after conducting studies by scientists of NGRI.

High Temperature (150°C) HGS

These mainly occur in Himalayan region except in Tatapani and Agnigundala area in peninsular region and in the Andaman and Nicobar Islands.

Intermediate Temperature (150°C-80°C) HGS

These are widely distributed and exist particularly in all geothermal regions of India. These form the common resource base in India.

Low Temperature (<80°C) HGS

These occur generally in limited numbers in almost all geological regions of India.

A Geopressured Geothermal System

In Combay basin. Low temperature (<80°C) hot water aquifer system. In some sedimentary basins.

Heat Energy Potential in India

Based on one hundred well recognised hydrothermal systems the estimated energy heat energy potential would be about 40×10^{18} calories which is equivalent to 27.6 billion barrels of petroleum and there would be a saving of 60,000 million rupees per year. The estimated power potential of the identified systems would be 2000 MW for utilization for a period of 30 years.

The studies carried out so far have indicated good prospects for developing Indian geothermal resources for generation of electricity by using binary cycle turbines and for non-electrical applications like space heating, poultry farming, mushroom raising, drying, mineral processing, etc., as mentioned already, cold storages could be developed too for preserving fruits and vegetables.

There are good prospects of tapping intermediate (80°C-150°C) and high temperature (>150°C) geothermal fluids at suitable locations in various parts of Himalayan region. Here very cold climate conditions prevail during most part of the year and the work season is very short, and the reserves of fossil fuels are meagre.

As regards peninsular India, geothermal resources of intermediate (150°C-80°C) and low grade (<80°C) at various locations are available and proposed uses have been already mentioned. Good possibilities do exist in peninsular India for developing industrial centres around geothermal fields. The geothermal fluids of these fields could be utilized for hydrothermal processing in ceramic, electronic, magnetic, pigment, inorganic chemicals, food and even iron/steel and non-ferrous metal industries. Hydrothermal processing involves the chemistry of hot water under pressure to carry out leaching and precipitation reactions and it has been emerging as a very potential process these days. It is hoped that several biotechnological units where biochemical reactions require large heat capacities but at lower temperatures, would be established in India also by them. Some of these units could be established near suitable geothermal fields.

Control of coal mine fires occuring mostly in Raniganj and Jharia coal fields, causing huge loss of prime coking and superior quality of non-coking coal, which cost Rs. 100 crore could be possible with the application of emerging technology of the down hole coaxial heat exchanger system. There is great possibility for the economic in-situ extraction both of the geothermal energy and the heat available from the combustion of coal from the coal fields of Jharia and use it for power generation.

By utilizing geothermal and non-conventional energy resources like solar, wind and biomass energy remote and isolated areas like Ladakh and many other parts of India could be economically developed. There is bright future for the utilization of geothermal energy resources on large-scale from socio-economic development in different parts of India.

Most of the Himalayan geothermal systems whose estimated aquifer temperatures range from 120°C to 240°C could be developed since the mean surface heat flow is higher than that of peninsular India. The thickness of soil crust in the Himalayas is more or less double that of peninsular India. It is also inferred that the natural heat discharge through the known geothermal systems of peninsular India and most of those in the Himalayas is derived from regional terrestrial heat flow on account of deep infiltration of meteoritic water upto a depth of 4 to 7 km. in the Himalayan region and 2 to 6 km. in peninsular region and circulation through deep permeable fractures. The process of heat transfer is greatly enhanced by the forced convection which is operative in most of the Indian geothermal systems. The geothermal systems in both peninsular India and Himalayan regions do not seem to have derived heat from cooling crustal magmatic rocks.

22

METHANE GANS FROM COAL SEAMS: A SOURCE OF FUEL ENERGY

Methane gas from coal seams is a good fuel resource and there is need to tap this gas on a large-scale to cater to the domestic and industrial needs all over the world. With the depletion of fossil fuel resources rapidly the significance of this gas has been very much realized by several countries like USA, Canada and Germany.

The presence of methane in coal-beds was recognized several hundred years back as a hazard for coal mining since it burns. Being lighter than air it tends to form a layer along the roof and cavities; posing a danger to the safety of mines. Disastrous fires occured due to this reason in many coal-mines of the world as well as in India causing several deaths.

The process of exploitation of this gas is environmentally attractive in comparison to mining and requires no manpower to work underground. After tapping methane from coal-beds, the coal mines are safely mined, which can reduce explosion hazards during mining.

Explorers recognized that coal seam methane (CSM) could be profitably exploited from the coal-beds beyond minimal depths or depths beyond the present limits of mining. Poduction of gas from such depths reduces hazards of methane leakage in mines and thereby keeps the mines safety. Moreover this gas is more environment friendly than the natural gas produced from oil fields.

The benefit of exploitation of this gas is twofold. It serves as an additional source of energy. Secondly, safety of mining coal is assured after its exploitation, since there cannot be any explosions in mines due to burning of this gas.

Coal Seam Methane

Production and Distribution

The first coal seam methane production was in USA in early seventies, followed by Canada in early eighties, coal methane is now being economically viable as unconventional source of energy in USA, Canada and several European countries like UK, Germany, France and Belgium.

Production of methane has phenomenally grown especially in USA in the last few years. From of dozen wells in 1980, there are 7000 wells at present producing coal seam methane from different coal fields of USA with a capacity of 1.95 to 2.75 bcf of methane per day. Commercial production of methane in USA is from two areas, i.e., Black Warrior and San Juan basins of Alabama and Colorado-New Mexico respectively. This methane resource in USA is about 800 tef. This estimate is for the coal up to a depth of 900 metres. Large quantities of coal exist beyond 900 metres. Since the gas content increases with depth the above estimate of resource in USA is considered conservative. Besides European countries, Japan and Australia have also started exploitation of this gas from coal seams. In Japan this gas was exploited from Sunagawa mine and in Australia from the mines located in New South Wales.

Formation

Coal seam methane is a biproduct of coalification process which is considered as a waste during mining. The process by which buried plant matrial evolved from peat to anthracite is termed coalification. As mentioned already, one of the biproducts of coalification is methane (CH_4). During early stages of coalification at temperatures <50°C biogenic methane is produced by microbial deposition of organic matters in the sediments. The gas is commonly termed as 'marsh gas'. With rapid subsidence and burial and under suitable geological conditions the biogenic gas can be trapped in shallow reservoirs. Thus, methane produced at temperatures <50°C with the increase in coal rank is called thermogenic methane which constitutes most of coal seam methane. During the coalification process from peat to anthracite about 6000 cubic feet of methane per ton of coal is generated.

Methane gas generated, is partly retained in the coal-bed. Gas produced in excess of the retention capacity of the coal-bed migrates away to the adjacent reservoir rocks, i.e., sandstones immediately above or below the coal-bed.

The maximum amount of methane is generated during the transition (about 150°C) from high volatile—A to low bituminous coal. As coal matures from volatible through low bituminous rank it may produce more methane than it can store. Additional methane may be naturally released from coal with increase in burial depth and gets trapped in adjacent reservoirs, if available, else remains in the coal-bed itself and forms a gassy field. The gas retained in coal is called coal seam or coal-bed methane. The generated methane is stored in the coal. Coal is a good reservoir and can hold two or three times more gas compared to a conventional sand reservoir because it has long surface area (over 1 billion square feet per ton of coal).

Retention of Methane in Coal

Methane is retained in coal-beds in four ways:

(i) as absorbed molecules on internal surfaces or absorbed in the molecular structure of the coal,

(ii) as gas held on matrix porosity (macro and micro porosity)

(iii) a free gas within the fracture network and

(iv) as gas dissolved in groundwater within coal-bed.

Retention of methane in coal-beds in dependent on the petrographic composition of coal. Vitrunite group of coals with more micro porosity than other groups has more retention capacity. Methane quantity increases with increasing quantity of coal. Nature of the original organic matter in the sediment also has the control on the quantity of methane generation. The capacity of coals to hold methane by absorption at various pressures is a function of coal rank (the higher the rank, the greater the internal surface area). Medium to low bituminous coal is an ideal rank for the methane to be present in large quantities.

As mentioned earlier, the generation of methane is linked with coal. As the rank is a function of temprature and time, i.e., maturation, it increases with depth. Thus, a high rank coal is likely to contain more methane than a low rank coal and the former is likely to contain more gas at a greater depth.

Output of Methane Generated

Methane generation is related directly to the degree of coalification (maturation of organic matter) process. Volume of methane generated increases with the rank of coal. While 1 ton of anthracite coal with highest rank generates 419 cubic metres of gas, 1 ton of earthy and dull brown coal generates only 68 cubic metres of gas. The hard semi anthracite coal, just lower by rank to anthracite generates 333 cubic metres of gas per 1 ton.

World Estimate of Reserves

Based on published estimates the coal seam methane reserves are 4000-7000 Tef.

Gas Production Technology

A coal seam methane well will typically find the formation to be saturated with water. Generally initial production rate of gas is low and increases with time as the reservoir is dehydrated. Rise of water-level in the well bore must be pumped off until the coal begins to release free gas.

Most of CSM wells are at low pressure and will require compression to deliver the gas to the consumer. And if the well is shut in for any significant time, the gas production problems may occur due to water loading.

Prospects of Coal Methane Gas in India

Very little data is available for methane content of Indian coals. Recently CMRI, Dhanbad, has carried out some work on more emission studies and absorbed gas studies of a few unmined coal-beds from bore hole samples. These studies confine to Jharia and Raniganj coalfields of Eastern India, Terap coalfield of N.E. India being mined by the Eastern and N.E. Collieries. Data on other coalfields have not been generated and it is not possible to estimate methane.

Most of coals are below the threshold value of their thermogenic gas generations. As such, the India coals are not gassy as those of USA. However, eastern coalfields particularly Raniganj and Jharia are more gassy with records of several mine disasters.

Studies reveal that estimated gas content of most of the coal seams being worked out in the country is less than 1 M^3 per ton. Only a few seams show appreciable gas content. The high values of methane content estimated in different virgin coal-beds are in Chasnala, Anilabad, Parbatpur and Satanala areas. The highest value of methane concentration noted in Parbatpur block Jharia coalfield is 14.93 M^3/ton at a depth of 795 m from surface. The regression analysis of the gas and depth of overburden showed an increase of 1.3 M^3/ton per 100 metres increase in depth. It is observed that methane content is generally low 2 M^3/ton except in Raniganj coalfields where in some areas, the value is 4.27 M^3/ton at a depth of 260 m and increasing with depth as mentioned above.

It, therefore appears that Coal Seam Methane prospects are good only in eastern coalfields, e.g., Raniganj and Jharia, Daltonganj, Bokaro and Giridih areas. In other areas prospects are fair to poor. However, from better coal rank it appears that Pench-Kanchan, north-eastern coalfields and Jammu coals hold prospects of methane.

Out of several coal basins only five coals basins are considered to be potential CSM exploitation. These basins are Raniganj, Jharia, Bokaro, Karanpura and West Bengal petroliferous basins.

Conclusion

Minimum gas resource of Gondwana coals of India could be 1 tcm and most likely to be 1.57 tcm. The gas can be exploited as a potential supplementary source of pipeline gas for domestic use and small-scale industries. The maximum density of gas resources is expected in eastern coalfields while moderate density is expected in the north-eastern tertiary coalfields as also in the Madhya Pradesh coalfields judging from chemical and petrographic compositions, rank and maturity of coals. These areas are recommended by CSM exploitation. The gas content, depth, coal rank, water content and total thickness of coal are the factors which determine the economic feasibility of producing methane in thin potential areas. Coal seam based methane is a source of energy in India for exploitation in future to partly meet the energy crisis.

PART III
DIET AND HEALTH INTERRELATIONSHIP

23

DIETARY FIBRE FOR HEALTH

Dietary fibre is the indigestable component of complex carbohydrates present in plant foods. For a long time the significance of dietary fibre in balanced diet has not been recognised by nutritionists. Only carbohydrates, fats, proteins, vitamins and trace minerals were considered to be essential in a balanced diet. Of late, the significance of dietary fibre in food has been recognised by nutrition scientists after carrying out considerable research studies. These studies have revealed that dietary fibre has prevention and curative effects of several diseases like colon cancer, heart diseases, diabetes, gall stones, obesity, diverticulitis, constipation and dental cavities. The changing pattern of these diseases in various countries especially with the rise of these diseases in developed countries prompted researchers to link these ailments with the changing dietary pattern of the people. Even though there are several causes for the incidence of these diseases, one factor seems to be the deficiency of the dietary fibre in their diet. The dietary pattern has very much changed with the economic development. People in developed countries and affluent urban societies even in developing countries whose per capita income is high, consume more saturated fats, proteins, refined sugars and vitamins. The high fat, protein rich diet which is mostly non-vegetarian, is the root cause for several ailments as mentioned above. This costly food is consumed by these higher income groups of people. On the other hand the poor nations of Africa and Asia consume mostly cheap vegetarian food and this has been proved to be more healthier than the non-vegetarian diet of different societies. In the developed countries, even though the communicable diseases could be controlled, the wrong eating pattern has given rise to the incidence of non-communicable diseases due to absence of dietary fibre in their diet. Diet predominantly consisting of items like beef and meat with the total absence of fruits and vegetables, does not consist any dietary fibre. Eating high fatty, high protein, non-vegetarian food and fast foods in posh hotels have became the fashion and a sign of modernism nowadays in many affluent urban societies particularly youth, giving up their traditional high fibre containing food.

Types of Dietary Fibres and Constituents

Considerable research work has been done in the significance of dietary fibre in food during the last 10-15 years. The researchers could identify and classify the components of dietary fibre and the role they play in the human metabolism. The research findings indicate that there are different types of dietary fibres, and each has different chemical and physical properties too. Besides, each has a physiological role to play in the human metabolism. To make things simpler, the components of dietary fibre have been classified into two categories as water soluble and water insoluble and the components of each category perform a different role. The major components of these two categories of dietary fibre are cellulose, hemicellulose, lignin, pectin, plant gums and musilages. All these have varying physical and chemical properties.

The dietary fibre in its natural in tact state poses a physiological obstacle to the ingestion and digestion of nutrients. By rendering food more solid due to the fibre content, both digestion and absorption are slowed down. This sets a limit to the energy intake and insulin responses and can be viewed as nature's way of preventing diseases and over nutrition.

The food stuffs rich in dietary fibre are cereals, pulses, vegetables like carrot, cabbage, fruits like apple, citrous, guava, etc.

Water soluble fraction of components is represented by cellulose, most hemicelluloses and legnins. These primarily increase stool bulk nd reduce transit time in gastrointestinal tract. These also reduce intra colon pressure and appear to play a beneficial role with respect to diverticular diseases. In addition to enhancing bulk legnins absorb organic substances such as bile and cholesterol from intestinal lumen and appear to play a role in the cholesterol lowering effect attributed to dietary fibre. Vegetables, wheat and most grain fibres are best sources of this water insoluble and by and large non-digestable fraction. The water soluble fraction on the other hand shows down gastric emptying and also the rate of absorption of many nutrients. The most important clinical role of these fibres is to slow the rate at which the carbohydrates are digested and absorbed. It also slows down glucose absorption. Fruits, oats, and dried legumes or beans are best sources of water soluble fibre. The balanced diet should include both soluble and insoluble fibres.

It is not quite correct to say that dietary fibre is not digestable since a certain percentage is digestable. The fibre is fermentable in varying degrees by microorganisms present in the large intenstine. The degree of degradation depends upon water holding capacity and the molecular structure comprising the fibre. It is reported that 36 per cent of wheat bran is degradable. Fibre rich in pectin but poor in legnin is more extensively degraded. Pectins mucilages and gums are completely degraded whereas cellulose is broken down partially. Pectin, mucilages, to a lesser degree hemicelluloses, have a higher water holding capacity. Typically a high water holding capacity is associated with greater fermentability of the fibre source. Pectin in particular is a good water binding agent.

Lignin has been proved to be an effective bile acid absorbent, thus facilitating its faecal excretion. Cellulose has little binding effect on bile acid forming.

Effect of Dietary Fibre in Curing Diseases

Cholesterol Reduction to Prevent Heart Diseases

High fibre diet especially rich in water soluble fibre like wheat bran, oats, dried pulses and beans and other vegetables and pectin from fruits lowers serum cholesterol levels by reducing transit time through intestinal tract. Further the fibre rich diets (especially those rich in pectins and gums) reduce reabsorption of unused bile acids, cholesterol, sterols, and fats and also facilitate their extinction. The bile acids are derived from the body's cholesterol pool. When they are excreted through faecal matter, the serum cholestrol is utilised for the synthesis of fresh bile. High serum choiesteral levels have been identified as one of the risk factors for cardio-vascular diseases. When serum cholesterol is used up the risk becomes less. The total cholesterol as well as LDL (bad cholesterol) level can be reduced by consuming more fibre as vegetables, fruits whole grains especially oats and beans. People eating less refined diets have been found to have lower level of cholesterol. The water soluble fibre pectin traps cholesterol, removes it from the body before it can cause trouble. Grape fruits pectin reportedly lowers cholesterol by 6-7 per cent within eight weeks. Since one per cent reduction in cholesterol causes two per cent drop in the risk of heart diseases it is rated quite significant. All fruits contain varying amounts of pectin. Oat bran appears to lower serum cholesterol in the same way as pectin rich fruits. Use of oats is worth considering for high risk subjects.

Crushed isabgol seeds can constitute a super concentrated form of fibre, but more than that they significantly lower blood cholesterol levels. Isabgol can be easily incorporated into foods such as bread, ready to eat cereals and snacks. Studies have shown that it could also reduce LDL (bad cholesterol) levels by 8 to 20 per cent. The scientists are not clear, that how this unique fibre lowers cholesterol and LDL levels so efficiently. The possible explanation is that it inhibits absorption of cholesterol from the gut into the body.

Diverticular Diseases

Diverticular diseases constitute a major cause of disability in the elderly, although they do bother people of all ages. approximately 5-10 per cent of population over 60 years suffers with one or the other diverticular diseases in the western world. Diverticular are pouch like outgrowths which develop in the weak areas of the bowel wall. If they are numerous and become inflamed, diverticulitis results. Diverticulitis was virtually unknown in the 19th century but its incidence (which runs parallel to constipation) has now increased remarkably in the developed countries. It has been known that vegetarians with high fibre intake are less prone (12 against 33 per cent in non-vegetarians) to diverticular diseases clearly demonstrating the protective role of dietary fibre. Many theories have been put forward to explain how the disorder develops. Small hard faeces undoubtedly features of a fibre deficient diet are associated with narrowing of the colon and the formation of closed segments in which pressure increases during defaecation. Increased dietary fibre, particularly water soluble type like wheat bran, provides symptomatic relief in diverticulosis.

Colon Cancer

Cancer of colon has been linked to non-vegetarian low fibre fat. The colon is the site of major transformation and modification of dietary fibre in humans. Diet rich in roughage produces bulkier stools which dilute concentration of carcinogens and other harmful substances in the colon lumen wherefrom the fibre ensures their speedy removal.

Since the dietary fibres increase transit time and decrease the time of release of ingested food in colon, the diet having less fibre is associated with colon cancer and the dietary fibre may play a major role in decreasing the risk of colon cancer.

Gall Stones

The incidence of gall stones (containing about 90 per cent of cholesterol) is significantly less amongst those who consume fibre rich diet. It has been experimentally proved that addition of bran (fibre) in the diet makes bile less saturated. Further, the incidence of experimentally induced gall stones in animals tends to be reduced if fibre rich food is supplied to them.

Constipation

The value of dietary fibre in easing constipation has been proved beyond doubt. The fibre is the single determinant of stool weight/mass which in turn is diversely related to the constipation–cellulose, hemicellulose and legnins are the components involved. Fibre diet prevents the movement of food in the gastrointestinal tract from being rapid (diarrhoea) or too slow (constipation). It also theorised that the increased volume and softness of the stools brought about by the water holding capacity of the fibre reduces pressure in the colon and eases straining defaecation, thus preventing haemorrhoids. Correction of fibre deficiency in the food is the natural way to ease constipation which bothers the aged in particular.

The dietary fibres hold water thus making stools soft, bulky readily eliminated. Therefore a high fibre intake prevents or reduces constipation.

Numerous fibre supplements are available which are

recommended as laxatives, these comprise Isabgol husk (psyllium) or wheat bran. Isabgol husk has capacity to absorb 25 times its weight of water. This eases constipation because the absorbed water forms a large bulk faciliating evacuation. Agar obtained from sea too has similar properties.

Control Diabetes

Dietery fibre seems to prevent diabetes. The American diabetes association recommended diets for diabetes which are low in fat and high in complex carbohydrates and fibre. The reason is diabetics are prone to high fat levels and heart disease. Low fat diets appear to reduce risks of both. In addition, many clinical studies have suggested that a high fibre, high carbohydrate diet may lead to significant weight reduction. Weight reduction itself is known to improve diabetic control. Pectin has recently been reported to slow the gastric releate of ingested food especially sugars. Thus, blood sugar level uses more slowly after a meal with the resultant lower insulin responses. Recent studies have confirmed that fibre free substances like apple juice produce post prandial insulin level than do quantities of intact apples. Further, lower blood sugar levels are reported after pectin rich meals. Theoretically therefore, pectin is likely to improve diabetic control by reducing the absorption rate of sugars. Recent studies have shown that the gum present in fenugreek (methi) seeds which contains 40 per cent is most effective in decreasing blood glucose and cholestoral levels as compared to other gums.

Reduction of Weight

Many studies revealed the efficiency of dietary fibre in controlling weight gain. The fibre in the diet by sheer bulk causes gastric distortion which leads to a feeling of fullness and in turn suppresses hunger and ultimately curtails food intake. Fibre as described earlier is known to cut down nutrition absorption including fats from the gut Pectin mucilages and gums retard gastric emptying. There is reduced desire to eat and by this, energy intake can be maintained with the required range. This phenomenon helps in keeping a check on obesity. Another reason to decrease obesity is high fibre diets which have somewhat low coefficients of digestibility. Thus the net energy released to the body is little less than that from other diets containing high proportions of animal foods.

Adverse Impact of Dietary Fibre

Consumption of too much dietary fibre is not good. It may lead to mineral deficiency. This is because of the preventing of minerals from absorption which results from faecal excretion of minerals. This is why vegetarians are deficient in trace minerals like zinc, copper, iron, calcium, phosphorus, magnesium. However, this is perhaps the only adverse effect of the fibre. This may create a nutritional deficiency when diets contain marginal levels of minerals.

The high fibre diets of Africa and Asia for example are associated with a high incidence of mineral deficiency particularly zinc and iron. Most experts feel that 40-50 grams of fibre per day would be safe and beneficial. Further we know that different fibres have different physiochemical characteristics as well as metabolic roles to play. The recommended fibre intake should come from a wide variety of fibre sources so that variety of dietary fibre are represented. The WHO recommends an intake of 22-23 grams for every 100 kcl. The National Institute of Nutrition, Hyderabad, on the other hand has recommended minimum of 40-50 grams of fibre per day. It is not good to increase fibre levels too much. Too much pectin is known to retard vitamin B 12 absorption. Dietary fibre in moderate quantities as recommended by nutrition experts is most essential to prevent and cure certain diseases and to keep one hale and healthy.

24

VEGETARIANISM: THE IDEAL CHOICE

Scentific investigations have firmly established that vegetarian diet is far better and ideal for one's health and environment protection.

The structure of the human body is more suitable for a vegetarian diet. The early man was only a fruit gatherer and survived; he did not know how to make weapons to kill animals. The sharp claws that characterise carnivores are absent in man who does not possess pointed teeth like that of the tiger. Our lower jaw can move both vertically and horizontally whereas it can move only vertically in the case of meat-eating animals. Our saliva is alkaline to digest vegetarian food such as cereals, pulses, fruits and vegetables, whereas it is acidic in the case of meat eating animals.

The total length of the intestines is four to five times more than our height, whereas they are equal in meat-eating animals. The carnivores have larger kidneys and liver and also secrete more bile to digest the flesh. In all respects the human body is so designed to handle vegetarian food.

A Myth

There is a misconception that meat eaters are more strong. This is a myth which is testified by the fact that there are giant animals such as the elephant and the rhinoceros among herbivores. There is another wrong notion that vegetarian food does not provide adequate protein. It seems that an acre of cereals could produce five times as much protein as an acre devoted to livestock. Legumes (beans and peas) can produce 10 times as much protein. Leafy vegetables, spinach for example, will provide 25 times as much protein from an acre of land set aside for livestock for meat production. Nutrition experts feel the combination of pulses, cereals along with fresh vegetables, milk and curd provide adequate protein content, and it is a balanced diet.

Vegetarian food provides a lot of dietary fibre which is need for the absorption of vitamins and minerals. Dietary fibre is known to prevent constipation and colon cancer and to reduce cholesterol in blood which causes coronary diseases. Colon cancer is more prevalent in Argentina and Uruguay where meat consumption is higher than in rest of South America.

On the other hand, non-vegetarian food consists of high cholesterol which is harmful to the heart. It does not contain any dietary fibre which is essential in the prevention of certain diseases. Meat takes longer time to go round the intestines and thus turns toxic and the bowel walls have sufficient time to absorb the toxins. It has been proved by resarchers that at the time of killing, an animal gets frightened and toxic elements are produced in its body. When a man consumes the meat contaminated with toxins, he is affected. The absence of dietary fibre which cleans the intestines is another disadvantage.

Contamination

The preparation process of meat and other non-vegetarian items also causes contamination, and consumption of such foods causes certain diseases. According to the Director, US Food and Drugs Administration (FDA), the animal food preparation industries use strong drugs and pesticides. Ms. Maneka Gandhi has stated that antibiotics are essential in the production of beef and other animals foods. According to her, about half antibiotics produced in the world are fed to food producing animals. Some farmers dip the animals in highly poisonous lindane to get rid lice and fleas. For fly control, a larvicide is added to the animal's feed. Such potent drugs and pesticides are never used in the preparation of plant foods.

Vegetarian diet is more economic, feeds more people and protects the environment. The quantities of grains and cereals fed to the livestock in the former USSR and U.S. are enough to feed the entire third world. Ms. Gandhi pointed out: 20 million tonnes of grain provides 2 million tonnes of animal protein. Seven kilograms of grain produce 1 kilogram of meat. One protein unit of soya bean or rice or wheat takes an energy factor between 2 and 10, whereas beef, eggs and mutton take from 10 to 78.

A beef cow must be fed on 21 pounds of plant protein in order to produce one pound of animal protein. This means a large amount of protein is wasted by feeding it to animals and then eating it—rather than using the plant food directly. One half of agricultural land is used for fodder crops.

A vegetarian can subsist just on 0.72 acres of land a year while 1.63 acres are needed to produce flesh food. It is a misconception that there will not be sufficient foodgrains if non-vegetarians become vegetarians.

Non-vegetarian food production is causing environment degradation on a large scale. South American forests have been cut down to grow cattle for hamburgers.

According to Ms. Gandhi, each goat or sheep eats grains equivalent to four hectares each year. In its life time of five years it destroys 20 hectares of government land, before it is killed. It costs the government nearly Rs. 25,000 to restore the degraded land. As a result of this grazing, forest lands, lands on hill slopes and on the road sides village panchayat lands and government land is totally destroyed.

Beause of environmental degradtion water level has fallen in Haryana, the lower Sivalik hills becoming barren tanks to grazing of the goats, and, streams coming into Haryana drying. All the Project Tiger reas are on the verge of extinction. The best example is the Bharatpur sanctuary, because of the huge inflow of cattle and goats. Seventy per cent of all plantations of forest department are destroyed by animals which eat young plants. The government requires Rs. 6,000 to regreen one hectare. It is difficult for the government to spare such huge amounts.

Utilisation of Land

Large areas of agricultural land carved out of forest land to grow fodder for these meat-eating animals can be utilised to produce wheat for the poor. By exporting meat to West Asia we are in the same position as South America destroying our environment for the sake of other countries. To meet the damage caused by export of 1 kg of meat we have to spend five times the money we earn. The more we sell the poorer we become.

Although hunger is not a sequence of meat production, most of the world's 630 million hungry people are so poor that they cannot buy animal food and they have to depend upon grains. Feeding the hungry is not only feasible but can be greatly facilitated by change in the dietary habits among the meat eaters. In fact, the Americans have reduced the consumption of pork and beef and increased vegetables by 11 per cent and fruits by 7 per cent in recent years. Some outstanding personalities such as Leo Tolstoy, H.G. Wells, Albert Einstein, William Shakespeare, Julian Huxley, Isaac Newton, George Bernard Shaw, Benjamin Franklin, Charles Darwin, Plato, Socrates, Pythagoras, Emerson, Shelly and Mahatma Gandhi were vegetarians. To preserve health and environment, the people should prefer a vegetarian diet.

25

WALK YOUR WAY TO HEALTH

Significance and Kinds of Physical Exercise and advantage of Aerobic Exercise

Physical exercise is a must for everybody to keep fit and healthy. This is more essential for the aged and over-weights. Physical exercise is of two kinds. The first kind is known as aerobic exercise consisting of long period of activity during which more air i.e., Oxygen is inhaled. The word aerobic means air i.e, Oxygen consuming. In this category walking, jogging, running, swimming climbing, dancing, playing games like Tennis, Badminton, foot ball, etc., are included. The other kind of physical exercise is anaerobic exercise. In this category short period strenuous activities like weight lifting, jumping, etc., are included. These physical activities engaged muscle fibre for this work and it is Glycogen in the muscles which is mostly converted into energy. As such the anaerobic physical exercises draw energy from Glycogen only. On the other hand, the aerobic physical exercises have more energy directly drawn from body's fat storage fatty acid and ketone bodies and hence these exercises are scientifically proved to be more beneficial for physical fitness of a human being.

Reduction of Weight by Additional Energy Expenditure

One of the major benefits of walking is the reduction of excess fat from the body. As such it is most advantageous for over weight people. Walking needs additional energy which is drawn mostly from fat storage. Energy is required for all vital functions of the body like, respiration, heart functioning, blood circulation, etc., besides several other activities like walking, writing, talking, etc. This energy is derived by the complex mechanism of chemical and physical processes which are involved in the body building and its maintenance. This complex mechanism of physical and chemical processes whose functions are two fold, is known as metabolism. Since the generation of energy which is the major function of metabolism, it is referred to generally as energy supplier. The utilization of energy needed for all vital functions of the body at rest is known as basal metabolism. The rate at which the energy is utilized for all vital functions is called basal metabolic rate which is referred to in abbreviated letters as BMR and this depends upon ones age, weight and sex. When a persons is walking, he needs additional energy for this activity which is mostly derived from fat storage, the body's metabolic rate also increases. Muscles normally use 20 per cent oxygen when a person is at rest. But when he does walking, the proportion is increased by 35 per cent. Our body continues to take up this oxygen deficiency. This process can keep our metabolism at a higher rate long after we stopped exercise. The higher amount of energy consumption i.e., burning of calories of energy from fat and glycogen storage is done by additional supply of Oxygen which is generated by walking exercise. In this way walking would help in reducing one's body weight. Walking or any other aerobic exercise burns fat from all over the body and not from any specific areas. It is impossible to reduce the amount of fat in a particular area without affecting the amount in other parts of the body.

If a person wants to lose 1 kg. of body weight he has to spend 7500 calories of enery. If one burns 250 calories of energy every day he may loose 1 kg. of body weight in a month.

Calculation of Energy Expenditure

The energy spent by a person depends upon ones body weight and speed of walking. A British Scientist Mr. Pass More suggested the following linear regression equation for calculating the energy spent per minute according to one's body weight when the spead of walking is about 3 miles (4.8 kilometres) per hour. The amount of energy spent at other speeds can be computed from this figure.

$E = 1.02+0.047\ W$

E = Energy

W = Weight of the person in kgs.

1.02 = Constant

The amount of energy spent in one hour by a person of 70 kg. weight would be (1.02 + 0.047 x 70) x 60 = 258.6 calories.

The amount of energy spent can be also calculated from the tables. The following table as suggested by Pass More and Durham can be used for calculating the energy expenditure under different body weights and speeds. The figures are multiplied by 0.0144 to get the amount of energy spent per minute.

TABLE 1

AMOUNT OF ENERGY SPENT ACCORDING TO BODY WEIGHT (IN KGS.) AND SPEED

Speed	*Weight (in Kgs.)*					
m. ph.	*45*	*55*	*65*	*75*	*85*	*95*
2.0	150	185	200	225	250	275
2.5	185	251	240	275	300	335
3.0	215	250	285	315	360	390
3.5	150	290	325	360	410	450
4.0	285	325	365	415	460	510

From this table, the energy spent for one hour walking by a person of 65 kg weight, walking at the rate of 3 miles per hour would be 285 x 0.0144 x 60 = 246.24 calories. The relationship between energy expenditure (calories per minute) and speed of walking (miles per hour) and body weight (lbs) is also given by Pass More in the following table from which the energy expenditure can be directly known.

TABLE 2

AMOUNT OF ENERGY SPENT ACCORDING TO BODY WEIGHT (IN POUNDS) AND SPEED

Speed	*Weight (in Pounds)*						
m. ph.	*80*	*100*	*120*	*140*	*160*	*180*	*200*
2.0	1.9	2.2	2.6	2.9	3.2	3.5	3.8
2.5	2.3	2.7	3.1	3.5	3.8	4.2	4.5
3.0	2.7	3.1	3.6	4.0	4.4	4.8	5.3
3.5	3.1	3.6	4.2	4.6	5.0	5.4	6.1
4.0	3.5	4.1	4.7	5.2	5.8	6.4	7.0

A person of 140 lbs walking at 3 miles per hour would spend 4 x 60 = 240 calories of energy in one hour of walking.

We spend the same amount of energy whether we walk, jog or run the same distance since we are moving the same weight to the same distance. On the other hand if we walk or jog or run for the same period of time there would be variation in the amount of energy spent because the distance covered by jogging or running is more than the distance covered by walking during the same period of time. If we jog or run for 30 minutes, we spend more energy than the energy spent in walking for 30 minutes since the distance covered in jogging in 30 minutes is more. On the other hand, if we walk or jog or run for 1 kilometre the amount of energy spent would be the same i.e. a fixed distance for a constant weight involves no variation in energy expenditure irrespective of the time factor. While this is generally acepted, some scientists hold the view that brisk walking utilizes more energy to cover the same distance.

Working Plan

Who can Walk

A person of any age with normal health can walk. But persons with certain ailments relating to heart or high blood pressure, or diabetes should consult the doctor first and take advice.

When Should We Walk

The best time for walks are the mornings. In the mornings the amount of oxygen released into the air is very high, and one feels fresh and is less tiring. The air is less polluted in the morning. If morning is not possible, you can walk in the evening or after dinner. But morning walks are preferable. You walk alone or in company or with a dog. You can have purposeful walking to attend some services like shopping or going to Post Office or Bank, etc.

Where to Walk

Select your walking track preferably away from pollution and congestion. To avoid monotony you can take different routes and directions.

Walking shoes and dress

Shoes: A pair of well fitting shoes, which offer adequate cushioning, reinforcement in the back around of the heel and flexibility in the toe area should be issued. The last is needed to allow toes to point out about 45° while you are walking. The shoes should have adequate space for the toe and are not pointed to the toe area nor are they too tight. While buying walking shoes go in for a slightly larger size rather than tight fitting. Buy the shoes in evening when your feet are slightly swollen. In good walking shoes, the heal counter (the stiff cup behind the tender and above the sole) in firm. The arch is not supported but is not jagged by the sole structure. The sole is light and thinner than a shoe used by runners in order to accommodate the lack of lift in the walking stride as opposed to running. In any case avoid wearing chappals as they are not conducive to brisk walking.

Dress: Do not over dress. Wear shorts preferably trousers. Wear any clothing that is comfortable, loose fitting and capable of absorbing perspiration.

Watch: A watch with a large second hand or digital second may be used for measuring your pulse rate.

Warming up Before Commencement of Walking

Every exercise requires warming up before the start. It gets the circulation going and increases the temperature in the muscles. This is called limbering and makes it easier to move aerobically and reduces the chances of injury and enables the muscles to form fuel efficiently. Repetitive movements tend to tighten the leg muscles. Hence, stretching should be done to compensate the same. When stretching carefully pull or lean against the joint or muscle you are stretching. Hold a stretch for 6 to 30 seconds. Do not stretch beyond the feeling of comfort.

Technique of Walking

Gray Yanker, a famous specialist on walking has propounded techniques of walking and the same are mentioned below:

Technique No. 1 is the holding posture while walking. The head should be erect, chin tucked in, shoulder back, buttocks and stomach are tucked in. However, shoulders should be relaxed. Overall, it should be an upright, balanced, relaxed posture. The feet should be parallel and about shoulder wide apart. The ear, shoulder, knee and ankle should be in straight line. The body should not lean forward and you should not lead with the chin.

Technique No. 2 is maintainting parallel and synchronised movement. The arms and legs should have parallel movement and the same should be synchronised with breathing. Arms should have easy Pendulum flowing movement. Arms will naturally bend at 90° at elbow and you will get a shortened pendulum movement—a kind of pumping action.

Technique No. 3 is called heelstrike and placement. While walking correctly the heel of the outstretched leg should hit the ground first. The leg will be at about 45° angle and toes will point straight ahead. The advantage is that this avoids binding of legs (called Pronation) and helps in keeping balance and develops more rhythmic movement.

Technique No. 4 is called leg pull-toe push off. After hitting the ground with the heel, roll forward to toe on outer edges of foot and simultaneously push off with toe of the back foot. Pull the back leg which will be slightly bent. This technique will help to develop balanced muscles.

Technique No. 5 is stride stretch. At every step, try reaching out just a few inches more. The greater leg extension will give more exercise and lengthens your stride. People have a misconception that to be able to walk faster you must move your legs quicker. However, lengthening the stride—even by a few inches as explained above will materially help in increasing the walking speed.

Technique No. 6 is bent arm pumping. 90 per cent of the time the arms are just brushing the sides. The fists are loosely clenched and they swing chest high giving a pumping motion. The movement of arms swinging at counter point to the legs across the front of the body can account for as much as 7 per cent of energy expended in walking. This ensures good exercise for the upper body.

After Walking

Drink fluids preferably water or have light warm soup in winter. Do not take a heavy meal for atleast one hour after vigorous exercise.

Duration of walking

Start walking programme slowly. Increase the distance starting with 1 kilometre. You can walk as much as you can, without exerting took much and without feeling it to be a boredom.

Speed of Walking

Based on Age: The speed of walking should decrease with age. The time taken to walk for 2 miles (3.2 kilometres) may be as follows:

TABLE 2.3

Age Group (Years)	*Times in Minutes*
20-29	19
30-34	21
35-39	26
40-44	27
45-49	30
50-60 and above	34

Based on Pulse Rate: Another way of determining the speed of walking and the distance to be covered according to ones age is based on the pulse rate. While walking the pulse rate is increased. You should see that your pulse rate is within target zone and should not exceed the target zone. Target zone pulse rate is arrived at, as follows. Subtract your age from 220, which gives the maximum pulse rate. The target zone is called training range and it is 65 per cent to 85 per cent of the maximum pulse rate. Thus, if we consider a man of 40 years, his maximum pulse rate would be 220–40= 180 and the target zone would be 110 to 155. The pulse rate is measured on the inside of wrist about 2 inches above the base of the thumb. Put your index finger in the little depression. Pulse can be measured in any convenient manner you can measure for 6 seconds and multipley by 10.

The target zone or training effect is a stage at which maximum amount of oxygen is being pushed out from all parts of the body. About 65 per cent of men able to get into the target zone after brisk walking for about 1.5 kms.

Precautions

If you have suffered from serious illness like heart attack, or heart enlargement, or having high blood pressure or diabetes, or over weight by more than 20 kgs. or aged above 50, better consult a doctor before taking up walking. Don't strain yourself with too much strenuous walking. It may cause fatique, breathing difficulty or pain in legs. If you develop pain in the legs stop walking for a day or two. Please see that your pulse rate should not go beyond the target zone. Control walking accordingly. Walking should be gradually increased.

Lastly, continuity of walking is most important for lasting benefits. So make walking a part of your daily life to keep yourself hale and healthy.

26

WALK YOUR WAY TO WEIGHT LOSS

Physical exercise is a must for everybody to keep fit and healthy. This is more essential for the aged and overweights. It produces a general feeling of well-being and relief from stress and tension. and keeps one in a good mood.

Exercise also helps in reducing the quantity of fat indirectly. It causes a fall in the amount of insulin in the blood. Insulin is a harmone secreted by the pancreas. It is a key agent in regulating how much fat we absorb and store. Overweight people typically have high levels of insulin which is necessary to keep fat stored. Due to exercise there is a decrease in insulin, leading to a fall in body fat.

As mentioned already, exercise increases the body's metabolic rate. Our muscles use 20 per cent of the oxygen. When we do exercise and become more active, this proportion increases by 35 per cent. Our body continues to make up this oxygen deficiency. This process can keep our metabolism at a higher rate long after we stop exercising.

Another benefit is that exercise prevents loss of lean tissue built primarily of protein. This tissue prevents a fall in the basal metabolic rate particularly when we are on a diet. Additional muscle tissue is also built up with exercise. With an increase in the metabolic rate, there would be reduction in fat. Heart and lung functioning will improve.

As age advances, the basal metabolic rate falls. In other

words, an aged man needs less energy for the body's essential functions. If he continued to take the same kind and quantity of food, the surplus energy is stored as fat and he would put on weight. If he wants to maintain the same weight and be healty, he should exercise.

Secondly, as one grows older, one's body shrinks from top to bottom. The muscles beign to lose strength and elasticity. Heart and lung power begins to drop, joints and ligaments become stiff and the body becomes more injury-prone.

All these can be kept at bay to an extent through exercise. With proper exercise and dietary habits, one can look younger and slimmer. Exercise not only enhances heart and lung functions, it improves skin tone and helps the body absorb nutrients.

There is no necessity to take up separate exercises to burn fat from different parts of the body. There is a mistaken notion that to burn fat from the abdomen, separate exercise is required and that it cannot be reduced by walking or running or some other exercise. It is impossible to reduce the amount of fat in a particular area without affecting other parts of the body.

Exercise also expands blood vessels in the muscles thus enabling the blood to be pumped more easily through longer pipes and removes strain on the heart. In an inactive person, the vessels tend to become narrower and harder resulting in stress on the heart and rising lood pressure. Exercise lessens the strain on the hearth and removes tension. It stimulates the development of blood vessels in the muscle tissue including that of the heart.

Strenuous exercise may cause breathing and sleeping difficulties and can also produce fatigue. Breathing should become normal within a few minutes after the cessation of exercise.

Disadvantages of Vigorous Exercise

People engaged in vigorous exercises do not gain any significant advantage over, those engaged in moderate or low physical activity. Moderate or low intensity exercises improved cardiovascular fitness. A British study reported that those engaged in moderate exercise gained more health advantages than people involved in vigorous sports. There is mounting evidence that strenuous exercise is potentially hazardous even for healthy and young people.

Vigorous exercises like running and jogging may cause injuries and the knees, legs, hips and feet may be affected. Very strenuous exercise like running can also cause special problems in women. It may cause harmonal changes. It has been found that 50 per cent of competitive runners developed irregular menstrual cycles and in some menses stopped altogether leading to temporary menopause and infertility. On the other hand, moderate exercise will strengthen bones in women and prevent osteporosis.

Benefits of Walking

As a result of health hazards, many runners and joggers of both sexes have switched over to walking. Studies at the University of Minnesota conducted by a cardiologist, Arthur Leon, indicated that brisk walking everyday leads to weight loss and reduces the risk of heart disease. If a person does brisk walking for an hour he would burn not less than 200 calories of energy. One can do brisk walking for at least 20 to 30 minutes or 2 to 3 km every day.

In addition to the benefits already mentioned, brisk walking helps tone up the leg muscles. It strengthens the arms when we swing them while walking. It helps circulation of blood through the body. When we walk, our muscles in the legs move, and squeeze the veins. This forces the blood back to the heart and gives you a feeling of well-being.

On the other hand, when we are sitting, or standing there is not enough pressure at the end of the veins in our legs to push the blood back to the heart efficiently. Since the leg muscles are the longest in the body, their work is enormously important. But if they are not being used with any vigour, they do not squeeze the blood towards the heart with any force. Brisk walking positively affects the human capillary system.

There are numerous blood vessels (whose combined total length comes to 96,000 km) in the body, mostly capillaries. These minute vessels are responsible for irrigating the flesh. Only a few will open when a muscle is at rest. Perhaps fifty times as many cells will open when a muscle is being exercised. Walking also reduces fatigue and helps in shaping the body. It resists backache which usually occurs in lower backbone muscles in old age.

Therefore, walking should not be avoided. If we develop any pain in the legs and feet due to muscle strain, we could stop walking for a day or two and then slowly come back to the normal schedule.

If you walk, jog or run the same distance, you spend the same amount of energy. But if you are engaged in any of these activities for the same period of time, there would be a variation in the amount of energy burnt. We burn more calories while jogging compared to walking for the same period of time. Similarly, we burn more calories while running when compared to jogging over that period of time.

Our ancestors were firm believers in the values of walking. Doctors today suggest brisk walking for everyone. It is the easiest exercise for most individuals. It is the most natural exercise, simple and safe. If walking cannot be done due to old age, or due to any physical ailment, or due to bad weather, we can choose any other aerobic-exercise of low or modernate intensity. As far as possible exercise should not be avoided. It keeps a man hale and healthy throughout his life.

27

KNOW YOUR HEALTH RISK FROM YOUR WEIGHT

Overweight or obesity has become a burning problem in the present day modern society. People with overweight seem to be more in educated urban societies of middle of higher income groups who are mostly having sedentary occupations. Of late there has been much awareness and anxiety among these people to reduce their excess weight by various methods. Due to this growing concern several slimming aid clinics have come into existence, charging heavily. Besides these, a number of health clubs and health farms have also cropped up in developed countries like USA. There is also too much awareness among boys and girls even in India to look slim and smart and there are frequent visitors to these slimming aid clinics.

Overweight is simply called obesity which is also expressed as body fat. A person is considered to be obese if he / she has 20 per cent above the ideal or normal body weight. The ideal weight depends upon the age, sex and body build of the person. The life expectancy is consideres to be shorter in the obese people. Insurance company and other data suggests that the risk to life is increased as weight increases. It is believed that people are susceptible to a number of diseases like heart diseases, blood pressure, gall bladder diseases, diabetes, arthritis, backache, etc.; Obese women particularly develop menstrual problems. They are likely to have miscarriages and childbirth may be more difficult. Anaesthetic complications are common after operation. Obese women are prone to uterus and breast cancer diseases. They are also susceptible to develop kidney diseases.

Though obesity may cause many of these diseases, it is not considered to be having any direct effect on coronary diseases like heart attack. Instead blood pressure and other factors like high levels of cholesterol which are associated with overweight are considered to be linked with coronary heart diseases. So it is these factors which are the direct cause of death, but not the overweight itself. If an overweight person maintains normal blood pressure and cholesterol level he is not likely to get heart diseases. But the health risk is more in obese person and it increases with overweight. For instance the risk of getting diabetes may be thirty times more if the weight of a person is 45 per cent more than the normal weight.

To keep safe from health risk by reducing excess weight one should know the degree of his health risk as per his existing weight. This can be expressed by using the body mass index (BMI). The BMI can be calculated by dividing one's weight in kgs with his height in square metres, for example, if a man's weight is 70 kgs and his height is 1.5 metres his body mass index would be 70/(1.5) 2= 31.1. The health risk of a person goes up with the increase in body mass index. If the BMI is between 25 and 30, there is low health risk, with a BMI over 40 there is high risk of people with normal weight.

It is therefore necessary for every one to know his health risk position and get his weight reduced to normal level by changing his diet and doing exercise in order to keep his normal health. People with overweight seem to be higher in urban society particularly in higher income groups in our country. Most of these overweight people are anxious to reduce their excess weight by resorting to various methods. A number of crash diets have come up into the market. Several surgical methods like intestine by-pass surgery and suction lipoctomy have been also developed for fat removal. All these methods develop harmful side effects.

Overweight is expressed as body fat and is called obesity. A person is considered to be obese if he or she has 20 per cent above the normal weight. Overweight people usually have protruding belly and huge fat accumulated around their neck and other parts of their body and put up an ugly appearance. The ideal or normal weight of a person depends upon ones height, sex and body frame. It has been established that obese people are susceptible to a number of diseases like heart attack, diabetes, gall bladder stone, backache, etc. Obese women are likely to have miscarriges, uterus, breast cancer and kidney diseases.

In overweight person the cholesterol, a sticky substance derived from fat is present in larger proportion in blood and this may cause heart attacks.

The energy consumption for certain vital functions like heart beating, respiration and blood circulation is low in fat people and this process of energy consumption for vital functions is called basal metabolism. Due to low basal metabolic rate (BMR) the fatty people are inactive and sluggish. Secondly, in obese people there will be high level of insulin which leads to more storage of fat.

Basically obesity is caused by two factors, i.e., heredity and overeating of starchy and fatty foods. If both the parents have overweight, the chances of their children having excess weight will be over 90 per cent. Due to psychological defects also obesity is caused. Due to inadequate production of a substance called (1) Cholocystokinin (CCK) in a small region in the brain called hypothalamus, obesity is caused. The hypothalamus which controls appetite does not function properly in a obese person. As such, an obese person eats more and feels hungry always. Apart from this, the presence of a small substance called adenosine triphosphatase (ATPASE) may cause obesity. Eating at large intervals, lack of exercise, excess use of common salt which causes fluid retention, constipation and stagnation of bile secretion, and improper digestion are other contributory factors for obesity. There are certain psychological factors that induce hunger like: salination in response to sight and smell of food; secretion of more insulin in the presence of eating cues; stress and strain; eating as a learned response at a particular time whether one is hunger or not; arousal, places like kitchen; activities like seeing television; food storage in a refrigerator and more number of items in lunch or dinner act as stimuli. The response to these stimuli is more in obese people.

As mentioned already obesity is caused mostly by consuming more calories of energy which depends upon the type of food and quantity. The energy required for body's vital and other functions is derived from essential nutrients particularly carbohydrates, fats and proteins like: Cholecystokinin and Adenosine triphosphatase (ATPASE) from the food we eat. The unit of human body energy is called kilo calorie or simply called calories and is abbreviated as Kcal.

Among the food nutrients, fats give higher energy. One gram of fat gives 9 calories of energy whereas one gram of carbohydrates and one gram of proteins give four calories of energy each. The energy requirement of a person depends upon one's body weight, sex, age and activity. Light work done by an office assistant sitting in a chair requires less amount of energy when compared to moderate or heavy work done by an agricultural labourer. From the physiological constitution point of view, a woman requires less amount of energy when compared to a man. The energy requirement also decreases with age.

A person in the age group of 60 to 70 years requires 80 per cent of energy required by a person under 30 years of age.

For undertaking the weight reduction programme the first and foremost thing one should do is to find out how many kgs. of excess weight he or she is having above the normal or ideal weight as per one's height, sex and body frame. For this, he should first find out his actual weight on a weighting machine. Then he could find his normal weight as per his height and body frame from height-weight tables. If a person's actual weight is 70 kgs. and his normal weight is 65 kgs., the excess weight would be 5 kgs. Since 1 kg. of body weight is equivalent to 7500 calories of energy, the total energy stored in his body as fat for 5 kgs. of excess weight would be 7500 x 5 = 37,500 calories of energy. If he wants to reach his normal weight of 65 kgs. he should burn all these excess calories of energy within a span of time. One should also know the exact amount of energy required as per his/her body weight and age. For a person of light activity for every one kg. of body weight, 43.6 calories of energy for a man, and 42.2 calories of energy for a woman are required for the maintenance of body. But this requirement decreases with age. A person above 60 years needs only 80 per cent of energy needed by a person under 30 years of age. A person under 30 years of age engaged in light activity weighting 70 kgs. needs 70 x 43.6 = 3052 calories of energy for his body maintenance. If his weight decreases, the requirement of energy also decreases at the rate of 43.6 calories per kg. However, the requirement of energy per kg. body weight increases with the level of activity. Moderate and heavy activities require more energy per kg. of body-weight.

Having calculated the excess weight over normal weight, one should know a little about the nutrition value of all the foodstuffs he consumes. For our growth and survival, we need certain nutrients. These are carbohydrates, fats, proteins, vitamins and minerals. But we derive energy mostly from carbohydrates. Proteins and fats also provide energy. Carbohydrates are mostly obtained from cereals and vegetables, fats from milk products, meat, eggs and oil seeds, and proteins from milk products, pulses, fish, eggs, etc. Vitamins are obtained from most of the food stuffs, of both vegetarian and non-vegetarian. Minerals are available mostly in green vegetables. In general, cereals, pulses and oil seeds, meat, eggs, dairy products, like butter and ghee give more calories of energy. Among the vegetables, root vegetables like potatoes contain slightly moe number of calories. All green vegetables, skimmed milk products contain few calories.

Most of the dieticians in their books suggest drastic cut in food intake and recommend low calories menus to lose more kilos of weight in a short time. This procedure in rather difficult and impracticable. Firstly while dieting people find it difficult to cut the quantity of food to such low levels. Even if they follow for few days, they come back to normal position. Secondly, the diet menus mentioned in their books particularly written by foreigners do not suit Indians. Even the menus suggested by Indian dieticians in their books do not suit many since the food items vary from region to region in our country.

To vercome all these difficulties the obese people should plan to reduce their weight slowly in longer duration of time. One can plan to reduce 1 kg. of body-weight per month without any difficulty. It amounts to a reducatieon of 250 calories per day from his normal intake of energy the reduction of energy expenditure could be achieved partly by substitution of low calories food items, partly by reducing the quantity of food, and partly by increasing the physical activity like walking, gardening, playing a game like badminton, tennis, etc. One can cut down about 130 calories of energy by foregoing one cup of rice or two *chapthis* each day. The remaining calories he can burn through physical activity. By walking a kilometre distance, one can burn 40 to 50 calories of energy on an average.

Physical activity is a must in a weight reduction programme. If weight reduction is adopted only through dieting without any physical activity, it causes the loss of lean tissue which is made up of proteins. The lean tissue is not only important for muscle formation but it also prevents the fall in basal metabolic rate. Any physical activity or exercise raises basal metabolic rate by supplying more oxygen to the body muscles, etc. If a person does too much exercise he would spend more energy, thereby he would feel hungry and is likely to eat more.

As such, it is better to lose energy partly through dieting and partly through exercise.

Many dieticians suggest drastic cuts in diet and total avoidance of sweets, and sugar in coffee, and tea. It is not possible for anyone to avoid completely social functions like marriages, birthday parties where there is a likelihood of overindulgence in consuming high calorie food items like sweets, ice cream, etc. Any excess calories intake on such occasions can be compensated on the next day partly by consuming more low calorie food items and partly by increasing any physical activity like walking.

Weight Reduction Monthly Plan

As mentioned already, one should plan to reduce body-weight by one kg. per month by cutting down 250 calories of energy per day. When he loses 1 kg of weight at the end of first month, he should start cutting down 300 calories of energy per day for reducing another 1 kg. of body-weight by the end of second month. In other words, a person of 70 kg weight after reacting 69 kgs. at weight at the end of first month, should start cutting 300 calories of energy per day, because the normal energy requirement in the case of a person of light activity falls by 43.6 calories per kg. of body weight. This could be taken as round number of 50 calories for every kg. loss of weight. It means that the normal energy requirement of a person of 68 kgs. weight is shorter by 50 calories of that of a person of 69 kgs. weight. For a person of 70 kgs. weight, whose normal weight is 65 kgs. he should cut down in the fifth month 250 + 50 + 50 + 50 + 50 = 450 calories of energy to reach his normal weight of 65 kgs. in five months. The energy expenditure could be proportionally increased partly through dieting and partly through exercise.

Precautions to be observed during weight reduction

There are few other important precautions that one should observe when one is on a weight reduction programme.

1. One should eat slowly doing a lot of mastication which involves muscle activity thereby burning up more energy. Masticated food reduces the risk of false appetite leading to obesity.
2. One should not eat at long intervals. This would minimise hunger and is tempted to consume more food. Three or four smaller meals spread through the day is most ideal and it will have far greater energy using effect.
3. Minimise the number of food items to avoid temptation of over-eating.
4. Have light dinner at night preferably with very low calorie food items since energy consumption is very low during sleep.
5. Drink enough water about 6 to 8 glasses. Water helps in metabolising the existing fat as well as in restricting the appetite. Furthermore, it counteracts the fluid retention.
6. Consume fibre rich foods like salads, green leafy vegetables, and fruits which increase the bowel activity and encourages bowel movements. The sluggish bowel activity upsets metabolism and the excess calories that are not burnt off, deposited as fat.
7. Decrease the consumption of salt. It causes water retention which prevents loss of weight.
8. Eat in a smaller plate so that you feel that you can take the same quantity twice, and you will have thorough satisfaction that you have eaten plenty.
9. Concentrate on food while eating; otherwise you are likely to eat more.
10. Shift your thought away from food. Keep your mind engaged in some activity like playing a game or reading a book so that your attention is diverted from the temptation of eating.

BIBLIOGRAPHY

1. Annual Report of Ministry of Non-Conventional Energy Sources, New Delhi, 1997-98.
2. 'Alaska Melting,' *Down to Earth*, Centre for Science and Environment Publication, 15th October, 1998.
3. 'Acid rain arriving soon in India'–Report, *Down to Earth* Journal, Centre for Science and Environment Publication, New Delhi, 15th May, 1999.
4. Anil Agarwal and Sunita Narayan, *Dying Wisdom* 'Rise and fall of India's traditional water harvesting systems,' Centre for Science and Environment Publication, New Delhi, 1997.
5. Bharadwaj Kumar Karthik, A.K., 'Hazardous wastes-Traffing and Disposal,' *Yojana*, November, New Delhi, 1997.
6. Bhakta Vatsalam, V. Chakravarthy, D., 'Non conventional energy for rural development'-*Development of India's resource base—Patterns, Problems and Prospects*, A seminar volume, Gian Publishing House, New Delhi, 1988.
7. Biswas, K.K., 'Prospects of coal bed methane in India'–*Indian Journal of Petroleum Geology*, Vol. 14, No. 2, December, 1995.
8. Biman Basu, 'Ozone hole gets curiouser', *Science Reporter*, Vol. No. 29, No.6, CSIR Publication, New Delhi, 1992.
9. Chhatwal, G.R., *Encyclopaedia of Environment and Management*, Vol. 102, Anmol Publications, New Delhi, 1997.
10. 'Capping toxic chemicals,' *Down to Earth*, Journal, Centre for Science and Environment Publication, New Delhi, 28th Feb. 1997.

11. Dhar, V.K., Kaul, V.R. and Bhat, C.L., 'The CFC Dilemma.' *Science Reporter*, CSIR Publication, June 1991.
12. David K. Miller, *Fitness a life time commitment*, Surjeet Publication, Delhi, 1989.
13. Davis, I.L., *Geographical variations in coastal development*. Longman Publishers, London, 1977.
14. Edward J. Calabresi, *Pollutants and High risk groups*, A Wiley Inter Science Publication, New York, 1978.
15. 'Emissions unlimited', Report, *Down to Earth* journal, 30th August, Centre for Science and Environment Publication, New Delhi, 1998.
16. Geological Survey of India Records, Vol. 115, Part 6, 1987.
17. 'Global Warming or Warning', *Science Express*, 9th February, CSIR Publication, New Delhi, 1992.
18. 'Fade to Black', *Down to Earth* journal, 31st August Centre for Science and Environment Publication, New Delhi, 1998.
19. Fritz Rummul—Geothermal energy—an alternative or additional energy source?-*Applied Geography and Development*, A biannual collection of Recent contributions, Vol. 25, Institute of Scientific Cooperation, Tubgin, Germany, 1985.
20. Global Warming—a serious threat, *Hindu*, 13th June, 1992.
21. 'Heavens Hellish layer', *Down to Earth* journal, 18th November, Centre for Science and Environment Publication, New Delhi, 1996.
22. 'Heat is on', *Down to Earth* journal, 30th September Centre for Science and Environment Publication, New Delhi, 1997.
23. 'India a dumping destination', *Down to Earth* journal, 11th August, New Delhi, 1998.
24. Judith Rodin, *Controlling your weight*. John Wiley & Sons, Toronto, 1983.
25. Kanwar, K.C. and Usha Kanwar, Shah, S., *Friendly fibres*, Science Report, CSIR Publication, New Delhi, June 1997.
26. 'Krill Kill'-*Down to Earth* journal-28th February 1999, Centre for Science and Environment Publication, New Delhi.

27. King, A.M., *Beaches and Coasts*, Edward Arnold Publishers Ltd., 1959.

28. Keneth wark, Gecil F. Warner, *Air pollution-Its Origin and Control*, A Don Donnelley Publishers, New York, 1976.

29. Maheswar Dayal, *Energy—Today and Tomorrow*, I.B. Publication GOI, New Delhi, 1989.

30. Mohan L. Gupta, *Geothermal Resources of India*: Geotectonic applications—present status and prospectives—Terrestial heat flow and geothermal energy in Asia.

31. Mukund Govind Rajan, 'The Ozone Challenge' - *Hindu,* 16th February, 1992.

32. Nobel Grove 'Air—An Atmosphere of Uncertainity'— *National Geographical Writer*, April 1997.

33. *Nutrition Composition of Indian Foods* – National Institute of Nutrition Publication, Hyderabad, 1989.

34. Pratapa Chandra Nair, K.O., Are they Oceans or Dumping Grounds?, *Hindu*, 5th May, 1991.

35. Patra T.C., Pandey, A.K., Dutta, and H.C., 'Potential areas of coal seam methane in India', *Journal of Geological Society of India*, Vol. 47.

36. Praveen Kumar, 'Global Warming may flood—Lakshadweep', *Science Express*, 15th May, 1960.

37. Rakesh Singh, 'Dietary Fibres', *Yojana*, September, 1997.

38. Pradipte Ghosh, Global Warming, Natures Revenge, *Science Reporter*, Vol. 29, No. 6, CSIR, Publication, New Delhi.

39. Passamore R and Durnin J.V.G.A., Human Energy Expenditure, *Physiological Review*, Vol. 35, The American Physiological Society, Washington, 1955.

40. Rashmi Mayur, 'Noise—A silent killer', *Science Express, Indian Express, supplement*, May, 1990.

41. Ravi Sharma, Hole of a problem', *Down to Earth* journal, 15th January, Centre for Science and Environment Publication, New Delhi, 1994.

42. Raja, S., 'Warning—Noise Festivals—Damage your ears', *Science Express*, *Indian Express, Supplement,* 17th October, 1989.

43. Ramanujam, M.P., 'Acid Rain', Science and Technology supplement, *Hindu*, 19th March, 1998.

44. Rakash Kabsham, 'Hot and anxious', *Down to Earth* journal, 15th August, Centre for Science and Environment Publication, New Delhi, 1996.

45. *The State of the World Environment*, 1989, UNEP Publication, Nairobi.

46. Sengal, S.P., 'Noise Pollution', *Survey of Environment*, 1991, Hindu Publication.

47. Seshadri, B., 'Coming to terms with Global Warming', *Hindu* Sunday, 26th May, 1991.

48. Sengupta, U.K., Depletion of Ozone—A Threat to Planets, *Science Reporter,* CSIR Publication, December, 1995.

49. Soapurno Benerjee, 'The Enemy within', *Down to Earth* journal, 15th July, Centre for Science and Environment Publication, New Delhi, 1996.

50. Seshadri, B., 'The Problem of Acid Rain', *Hindu*, 16th Feb. 1992.

51. Trivedi, P.R., *Enclyclopaedia of Environmental Sciences*, Vol. 1 to 6, Ashadweep Publishing House, New Delhi, 1992.

52. 'Tirade against toxins', *Down to Earth* journal, 31st May, Centre for Science and Environment Publication, New Delhi, 1997.

53. Unmer Kitty, A.N.D., *Science of the Oceans and Human Life*, N.B.T. Publication, 1985, New Delhi.

54. Vidyanath, V., 'Non-conventional energy resource atlas of Andhra Pradesh', unpublished project report, 1995, sponsored by the Ministry of Science and Technology (DST), GOI, New Delhi.

55. Vidyanath, V., 'Kolleru Tract—A Case Study for Environmental Determinism', *Journal of Landscape Systems*, Vol. 8, No. 1, June, 1985.

56. Vidyanath, V., On the evolution and morphology of Kolleru lake, *Indian Journal of Landscape Systems*, Vol. 3, No. 1 and 2, 1980.

Index